SpringerBriefs in Earth System Sciences

Series Editors

Gerrit Lohmann, Universität Bremen, Bremen, Germany

Justus Notholt, Institute of Environmental Physics, University of Bremen, Bremen, Germany

Jorge Rabassa, Labaratorio de Geomorfología y Cuaternar, CADIC-CONICET, Ushuaia, Tierra del Fuego, Argentina

Vikram Unnithan, Department of Earth and Space Sciences, Jacobs University Bremen, Bremen, Germany

W0079149

SpringerBriefs in Earth System Sciences present concise summaries of cutting-edge research and practical applications. The series focuses on interdisciplinary research linking the lithosphere, atmosphere, biosphere, cryosphere, and hydrosphere building the system earth. It publishes peer-reviewed monographs under the editorial supervision of an international advisory board with the aim to publish 8 to 12 weeks after acceptance. Featuring compact volumes of 50 to 125 pages (approx. 20,000—70,000 words), the series covers a range of content from professional to academic such as:

- A timely reports of state-of-the art analytical techniques
- bridges between new research results
- snapshots of hot and/or emerging topics
- literature reviews
- in-depth case studies

Briefs are published as part of Springer's eBook collection, with millions of users worldwide. In addition, Briefs are available for individual print and electronic purchase. Briefs are characterized by fast, global electronic dissemination, standard publishing contracts, easy-to-use manuscript preparation and formatting guidelines, and expedited production schedules.

Both solicited and unsolicited manuscripts are considered for publication in this series.

Erik Velasco · Armando Retama ·
Dimitris Stratoulias

Air Quality Management and Research in Southeast Asia

 Springer

Erik Velasco ⓘ
Molina Center for Energy
and the Environment
Boston, MA, USA

Armando Retama ⓘ
Molina Center for Energy
and the Environment
Boston, MA, USA

Dimitris Stratoulias ⓘ
Asian Disaster Preparedness Center
Bangkok, Thailand

ISSN 2191-589X   ISSN 2191-5903 (electronic)
SpringerBriefs in Earth System Sciences
ISBN 978-3-031-69087-7   ISBN 978-3-031-69088-4 (eBook)
https://doi.org/10.1007/978-3-031-69088-4

This book is dedicated to everyone in Southeast Asia working for a sustainable, equitable and inclusive society in harmony with the environment.

Foreword

Air pollution has become one of the most important environmental challenges. Rapid population growth, uncontrolled urban expansion, unsustained economic growth, and increased energy demand have led to the emission of large quantity of harmful pollutants and greenhouse gases into the atmosphere. Reports from World Health Organization and other public health institutions show that air pollution continues to pose a significant threat to global health; it is the greatest external threat to human life expectancy on the planet. These reports highlight the unequal distribution of the pollution burden throughout the world and the basic resources needed to design cost-effective pollution control policies. Although some regions around the world have achieved substantial progress in tackling air pollution, many cities are still struggling with environmental degradation. Air pollution levels remain dangerously high in many regions of the world, particularly in the low- and middle-income nations.

This monograph is the first book that provides a comprehensive and critical review of the current state of air quality management and scientific research in Southeast Asia, with the aim of initiating a science-based conversation to address the air pollution problem and implement effective emission control strategies in the region. It is written by three experts in air pollution science who have conducted extensive research on the subject and have first-hand knowledge of the challenges facing the region.

Southeast Asia region comprises of eleven nations: Brunei Darussalam, Cambodia, Indonesia, Lao PDR, Malaysia, Myanmar, Philippines, Singapore, Thailand, Timor-Leste, and Vietnam, with distinct geographical features and climates across the region. The region has a population of approximately 675 million (in 2021); it is one of the most culturally diverse regions of the world, with many different languages, ethnicities, and religions. The income levels are also diverse, except Singapore and Brunei, the other countries are classified as low income and middle income. The Association of Southeast Asian Nations (ASEAN) was established to promote intergovernmental cooperation and facilitate economic, political, security, educational, and cultural integration among its eleven members (East Timor is an observer state). The region's diversity in terms of culture, politics, geography, and economics presents some of the governance and social challenges that the ASEAN member countries face in their highly diverse and unequal societies.

As in many other regions around the world, air pollution is worsening in Southeast Asia as a result of rapid population growth, uncontrolled urban expansion, and increased energy consumption. Air pollution is responsible for 335-thousand premature deaths in the region each year, incurring a monetary cost equivalent to 6.75% of the regional GDP. In order to design effective emission control strategies to protect public health, it is essential to have the tools that enable timely and reliable air quality management based on scientific evidence derived from accurate data sources. Additionally, national air quality policies should be extended to a regional scale so that all nations can jointly implement effective emission control measures that benefit all sectors and actors. However, as noted in the monograph, the knowledge on air pollution is limited and insufficient to take informed action at the local and regional scales.

The material presented in this monograph is the result of critical review of the available data derived from air quality management and scientific research in the region by the authors, providing comprehensive information on urban and regional air quality issues and their implementation within monitoring, emissions inventory, modeling, risk assessment, and management approaches. It also highlights topics that have not yet been adequately investigated, such as atmospheric chemistry in a tropical climate and the latest technologies used in air quality assessment and monitoring of pollutants. It is a valuable resource for researchers, policymakers, environmentalists, stakeholders, as well as students and the general public who are interested in the state of air quality in Southeast Asia.

This monograph provides an invaluable and timely contribution to the urgent environmental challenges facing Southeast Asia. In presenting what is known about the causes and consequences of air pollution in this region, the authors highlight what needs to be done and include a set of recommendations for developing a roadmap to address the current gaps in air quality management in Southeast Asia based on the region's specific conditions and needs. Concerted efforts among air quality managers, policymakers, and the scientific community from all countries in the region are needed to meet the challenges and achieve clean air.

Boston, MA, USA   Luisa T. Molina
May 2024

Preface

The threat of air pollution requires concerted efforts at local and regional levels. In regions like Southeast Asia, each nation is an emitter and a receiver of atmospheric pollutants. Therefore, national air quality policies should be extended to a regional scale so all nations can jointly implement effective emission control measures that benefit all sectors and actors. This must be done following a science-policy approach, in which decisions are made based on scientific evidence derived from accurate data sources. To accomplish this, it is essential to have the tools that enable timely and reliable air quality management.

Air quality management entails collecting data to characterize the magnitude, origin, and impact of air pollution to support corrective actions. Ambient air quality monitoring, emission inventories, and chemical-transport models, always supported by scientific research, are the basic tools for tracking the spatial and temporal distribution and evolution of air pollutants, identifying their origin, and assessing their chemical and physical transformations, dispersion, and fate discharge. This information is essential to account for the costs and consequences of air pollution on public health, social and economic development, livability, and environmental sustainability.

Unfortunately, air quality management in Southeast Asia has not yet been consolidated, so there are gaps in information and knowledge about the state of air quality. This is the main outcome of a comprehensive review of the available data derived from the air quality management cycle and scientific research in the region, as presented here.

This work examines the capabilities for monitoring air quality, developing emission inventories, and applying chemical-transport models for regulatory, forecasting, and research purposes throughout the Southeast Asian region. It also examines the scientific efforts made to better understand and characterize atmospheric pollution and its impacts on public health and highlights topics that have not yet been adequately investigated, such as atmospheric chemistry in a tropical climate, or the use of satellite remote products to assess the regional transport of pollutants.

The monograph was prepared to serve as a bibliographic resource for those involved in air quality management, including practitioners, researchers, and stakeholders, as well as students and members of the general public who are interested in the state of air quality in Southeast Asia. Readers will find information to understand the different aspects of air pollution, as well as references to delve deeper into particular topics. The monograph also includes a set of recommendations for developing a roadmap to address the current gaps in air quality management in Southeast Asia based on the region's specific conditions and needs.

Singapore, Singapore   Erik Velasco
Mexico City, Mexico   Armando Retama
Bangkok, Thailand   Dimitris Stratoulias

Contents

About the Authors

Erik Velasco is a research scientist with over 20 years of experience doing applied research in atmospheric sciences. He has led studies and participated in collaborative field campaigns to investigate the impact of urbanization and climate change on air quality, micrometeorology, and the emission of greenhouse gases. His findings have contributed to air quality and climate change mitigation programs, as well as improving public transportation plans and efforts to reduce urban warming. He studied mechanical and environmental engineering at the National Autonomous University of Mexico and received his Ph.D. from Washington State University. He has worked for research institutions in Mexico, the United States, and Singapore.

Armando Retama has over 30 years of experience as an air quality monitoring specialist. He has contributed to improve air quality monitoring in Mexico and Latin America. For 16 years, he oversaw Mexico City's air quality monitoring network. He consolidated the network and made it the backbone of the city's environmental management to address the threat posed by air pollution, becoming it one of Latin America's most important and reliable networks. He also conducts research on the subject, specializing in atmospheric chemistry. He studied chemistry at the National Autonomous University of Mexico.

Dimitris Stratoulias is a senior scientist and data expert with more than 15 years of experience in academic and non-profit organizations. He is leading evidence-based R&D efforts and guiding geospatial information services for Earth Observation programs through the development of data streams, machine learning algorithms, real time information systems, science, and advocacy. He holds a Ph.D. degree in remote sensing from the University of Leicester awarded with a Marie Curie fellowship, a M.Sc. in GIS from the University of Edinburgh, and a B.Sc. in Physics from the University of Athens. He has been living in 10 countries across Europe and Asia.

Abbreviations

7-SEAS	7-South-East Asian Studies
ACSM	Aerosol Chemical Speciation Monitor
ADPC	Asian Disaster Preparedness Center
AERONET	AErosol RObotic NETwork
AHI	Advanced Himawari Imager
AirNow-DOS	US Department of State's AirNow program
AIRS	Atmospheric Infrared Sounder
AIS	Acute ischemic strokes
AOD	Aerosol optical depth
APSI	Air Pollutant Standard Index
AQG	Air quality guideline
AQI	Air quality index
AQMS	Philippine's Air Quality Management Section
ASA	Active surface area
ASEAN	Association of Southeast Asian Nations
ASIA-AQ	Airborne and Satellite Investigation of Asian Air Quality
BaP	Benzo[a]pyrene
BASE-ASIA	Biomass-burning Aerosols in South-East Asia: Smoke Impact Assessment field campaign
BC	Black carbon
BenMAP-CE	Environmental Benefits Mapping and Analysis Program
BMKG	Indonesia's Meteorology, Climatology and Geophysics Agency
BrC	Brown carbon
BrO	Bromine
BTEX	Benzene, toluene, ethylbenzene, and xylene isomers
BVOCs	Biogenic volatile organic compounds
C_2H_4O	Acetaldehyde
C_4H_6	Butyne
C_6H_6	Benzene
Ca_2^+	Calcium ion

CALIOP	Cloud-Aerosol Lidar and Infrared Pathfinder Satellite Observation
CAMP2Ex	Clouds, Aerosol Monsoon Processes Philippines Experiment
CAMS	Copernicus Atmospheric Monitoring Service
CCl_4	Carbon tetrachloride
CEDS	Community Emissions Data Set
CESM2	Community Earth System Model version 2
CFCs	Chlorofluorocarbons
CH_3Br	Bromomethane or methyl bromide
CH_3CCl_2	1,1-Dichloroethane
CH_4	Methane
CHOCHO	Glyoxal
Cl^-	Chloride ion
CMAQ	Community Multiscale Air Quality model
CO	Carbon monoxide
CO_2	Carbon dioxide
COPD	Chronic obstructive pulmonary disease
COVID-19	Severe acute respiratory syndrome-coronavirus 2 or SARS-CoV-2
CrIs	Cross-track Infrared Sounder
DEPR	Brunei Darussalam's Department of Environment, Parks and Recreation
DKI Jakarta	Environmental Agency of Jakarta
DOE-ME&W	Malaysia's Department of Environment, Ministry of Environment and Water
DU	Dobson units
EANET	Acid Deposition Monitoring Network in East Asia
eBC	Equivalent black carbon
EC	Elemental carbon
ECLIPSE	Evaluating the Climate and Air Quality Impacts of Short-Lived Pollutants
ECMWF	European Centre for Medium-Range Weather Forecasts
EDGAR	Emissions Database for Global Atmospheric Research
ENSO	*El Niño*-Southern Oscillation
EPCI	Energy Policy Institute at the University of Chicago
ESA	European Space Agency
F^-	Fluoride anion
FPG	Fasting plasma glucose
FVM-TAPOM	Finite Volume Model-Transport and Photochemistry Mesoscale model
GAFIS	Global Air Quality Forecasting and Information System
GAIN	Green-house gas—Air pollution Interactions and Synergies model
GAW	Global Atmospheric Watch

GBD	Global Burden of Diseases
GC-FID	Gas chromatography-flame ionization
GC-MS	Gas chromatography-mass spectrometry
GDP	Gross domestic product
GEMS	Geostationary Environment Monitoring Spectrometer
GEO-KOMPSAT 2B	Geostationary Korea Multi-Purpose Satellite 2B
GEOS-CF	Goddard Earth Observing System-Composition Forecasting
GFAS	Global Fire Assimilation System
GFED	Global Fire Emissions Database
GHG	Greenhouse gases
GLOMAP	Global Model of Aerosol Processes
GMAO	NASA's Global Modeling and Assimilation Office
GOME-2	Global Ozone Monitoring Experiment-2
GOSAT	Greenhouse gases Observing SATellite
HCFCs	Hydrochlorofluorocarbons
HCHO	Formaldehyde
HCl	Hydrochloric acid
HCN	Hydrogen cyanide
HCO_3^-	Bicarbonate ion
HEI	Health Effects Institute
HFC	Hydrofluorocarbon
Hg	Mercury
HNO_3	Nitric acid
HO_2	Hydroperoxyl radical
HPLC	High-performance liquid chromatography
HULIS-C	Carbon content of HULIS
HYSPLIT	Hybrid Single Particle Lagrangian Integrated Trajectory model
IASI	Infrared Atmospheric Sounding Interferometer
ICP-MS	Inductively coupled plasma mass spectrometry
IFS	Integrated Forecasting System
IGAC	International Global Atmospheric Chemistry Project
IHME	Institute for Health Metrics and Evaluation
IOD	Indian Ocean Dipole
IT	Interim target
IVE	US-EPA International Vehicle Emissions model
JPSS	Joint Polar Satellite System
K^+	Potassium ion
$K_\downarrow$	Incoming shortwave radiation
KLHK	Indonesia's Ministry of Environment and Forestry
LEO	Low Earth orbit
LightGBM	Light Gradient Boosting Machine algorithm
MACE-2015	Manila Aerosol Characterization Experiment—2015
MAQE	Mekong Air Quality Explorer

MARGA	Monitor for AeRosols and Gases
MEGAN	Model of Emissions of Gases and Aerosol from Nature
MERRA-2	Modern-Era Retrospective analysis for Research and Applications, version 2
MetUM	UK Met Office Unified Model
Mg^{2+}	Magnesium ion
MICS-Asia	Model Inter-Comparison Study for Asia
MISR	Multiangle Imaging Spectro Radiometer
MIXv2	Mosaic Asian Inventory
MODIS	Moderate Resolution Imaging Spectroradiometer
MOPPITT	Measurements Of Pollution In The Troposphere
MOZART	Model for OZone and Related chemical Tracers
MSA	Methanesulfonate
N_2O	Nitrous oxide
Na^+	Sodium ion
NAAPS	Navy Aerosol and Analysis System
NAME	Numerical Atmospheric-dispersion Modeling Environment
NASA	US National Aeronautics and Space Administration
NCAR	US National Center for Atmospheric Research
NEA	Singapore's National Environmental Agency
NF_3	Nitrogen trifluoride
NH_3	Ammonia
NH_4^+	Ammonium ion
NMHCs	Non-methane hydrocarbons
NMVOCs	Non-methane VOCs
NO	Nitric oxide
NO_2	Nitrogen dioxide
NO_3^-	Nitrate ion
NOAA	US National Oceanic and Atmospheric Administration
NO_X	Nitrogen oxides
NRCM-Chem	Nested Regional Climate Model with Chemistry
O_3	Ozone
OC	Organic carbon
OCPs	Organochlorinated pesticides
OH	Hydroxyl radical
OHCA	Out-of-hospital cardiac arrest
OMI	Ozone Monitoring Instrument
P	Atmospheric pressure
PAHs	Polycyclic aromatic hydrocarbons
PAN	Peroxyacetyl nitrate
PAX	Photoacoustic Extinctiometer
Pb	Lead
PCBs	Polychlorinated biphenyls
PEFRs	Peak expiratory flow rates
PFCs	Perfluorochemicals

PM	Particulate Matter
PM_1	Particle matter with aerodynamic diameter equal or less than 1 μm
PM_{10}	Particle matter with aerodynamic diameter equal or less than 10 μm
$PM_{2.5}$	Particle matter with aerodynamic diameter equal or less than 2.5 μm
PMF	Positive matrix factorization
PO_4^{3-}	Phosphate ion
pPAHs	Particles bound PAHs
PPP	Purchasing power parity
PSI	Pollutant standard index
PTR-MS	Proton-transfer reaction mass spectrometry
PV	Photovoltaic
QA/QC	Quality assurance and quality control
REAS	Regional Emission Inventory in Asia
REMO	3D regional atmospheric chemistry model REgional Model
RH	Relative humidity
ROTRAJ	Reading Offline Trajectory Lagrangian 3D model
SF_6	Sulfur hexafluoride
SILAM	System for Integrated modeling of Atmospheric coMposition
SLCP	Short-lived climate pollutant
S-NPP	Suomi National Polar-orbiting Partnership satellite
SO_2	Sulfur dioxide
SO_2F_2	Sulfuryl fluoride
SO_4^{2-}	Sulfate ion
SPARTAN	Surface Particulate Matter Network
T	Ambient temperature
TANSO-CAI	Thermal and Near-Infrared Sensor for Carbon Observation-Cloud and Aerosol Imager
TANSO-FTS	Thermal and Near-Infrared Sensor for Carbon Observation-Fourier Transform Spectrometer
TAPM-CTM	Air Pollution Chemical Transport Model-Commonwealth Scientific and Research Organization
TEMPO	NASA's Tropospheric Emissions: Monitoring of Pollution mission
TOAR	Tropospheric Ozone Assessment Report
TROPOMI	TROPOspheric Monitoring Instrument
TSP	Total suspended particles
UFP	Ultrafine particles
UNEP	United Nations Environment Programme
US-EPA	United States-Environmental Protection Agency
VCD	Vertical column density
VOCs	Volatile organic compounds

WACCM	Whole Atmosphere Community Climate Model
w_d	Wind direction
WHO	World Health Organization
WRF	Weather Research and Forecasting model
WRF-Chem	Chemical module of WRF
w_s	Wind speed
WSOC	Water soluble organic carbon

List of Figures

List of Tables

Chapter 1
Introduction

Abstract Air pollution threatens public health and has a negative impact on people's lives and well-being. Above this, its impact on the environmental sustainability of natural ecosystems and agriculture, as well as its connections to climate change must be considered. This chapter starts off with a concise description of all of adverse effects associated with poor air quality, as well as the social and economic costs arising from them. They explain why governments must prioritize air pollution control on their agendas, as well as the need for accurate air quality information to design effective environmental policies. To attain clean air, it is necessary to collect reliable data to characterize the magnitude, origin, and impact of air pollution to support corrective actions. This is achieved through air quality management, the set of activities that regulatory authorities undertake to help protect human health and the environment from the harmful effects of air pollution. This chapter describes the air quality management process as an introduction to looking into and assessing the current state of air quality management and scientific research in Southeast Asia. The chapter also gives an overview of the monograph, outlining its organization and what the reader will find in each chapter.

Keywords Air quality management · Public health · Air pollution impact · Air pollution cost

Air pollution is an invisible threat to public health that increases mortality and reduces life expectancy by several months to a few years (Boogaard et al. 2019; Apte et al. 2018; Capello and Gaddi 2018; Brunekreef and Holgate 2002). Studies have found that people living in places with polluted air are more likely to experience health problems in the long term, including respiratory and cardiovascular illnesses, diabetes, lung cancer, and neurological disorders, as well as common mental health conditions, such as depression and anxiety (Ronaldson et al. 2022; Shi et al. 2022; Paul et al. 2020; Turner et al. 2020; Pun et al. 2017; Calderón-Garcidueñas et al. 2015; Franklin et al. 2015; Hoek et al. 2013; Brook et al. 2010). But also, exposure over a few hours to weeks can trigger cardiovascular and respiratory symptoms, especially in susceptible people, such as the elderly and those with preexisting medical conditions

© The Author(s), under exclusive license to Springer Nature Switzerland AG 2024
E. Velasco et al., *Air Quality Management and Research in Southeast Asia*,
SpringerBriefs in Earth System Sciences, https://doi.org/10.1007/978-3-031-69088-4_1

(Lu et al. 2022; Liu et al. 2021; Sinharay et al. 2018; Shah et al. 2015; Link et al. 2013; Wellenius et al. 2012; Atkinson et al. 2001).

The adverse effects of air pollution are more severe in vulnerable groups: children, the elderly, and pregnant women. Children and teenagers are in a state of development, and therefore, the influence of external agents in the air they breathe becomes more acute, affecting their growth and brain development, and increasing the risk of chronic diseases in their adulthood (Landrigan et al. 2019; Payne-Sturges et al. 2019; Capello and Pili 2018; Goldizen et al. 2016). In the case of older adults, the immune system no longer responds as efficiently, so the alterations caused by air pollution increase susceptibility to various diseases (Yap et al. 2019; Sabatini 2018; Simoni et al. 2015). Similarly, exposure to air pollution during pregnancy can cause problems for both mother and child. The gestational period can be shortened and the risk of miscarriage increases (Ha et al. 2018; Wu et al. 2016). The development of the embryo or fetus is altered, which can cause congenital anomalies, development and cognitive delays, lung problems, and alterations of the immune system in later stages of life (Johnson et al. 2021; Ha et al. 2019; La Marca and Gava 2018; Lertxundi et al. 2015). Likewise, the mother can develop hypertensive disorders and gestational diabetes (Hu et al. 2020).

Some medications lose their effectiveness when the patient breathes polluted air. Both drugs and pollutants are metabolized by the same enzymatic systems, which may reach a point where they can no longer remove xenobiotic compounds (Tumiatti et al. 2018; Le Vee et al. 2015; Fardel et al. 2012). And even worse, the chemical reactions that are responsible for detoxification of the body can begin to transform non-toxic xenobiotic compounds into toxic ones, exacerbating the negative effects of air pollution.

Air pollution affects people's lives and places a significant burden on economies and health services from a societal standpoint. Those who become ill as a result of breathing polluted air increase hospital admissions. People are less productive and more likely to take days off when they are sick, undermining their contribution to economic growth (McGrath et al. 2022; Chen and Chen 2021; Hanna and Oliva 2015; Graff Zivin and Neidell 2012). Air pollution causes learning difficulties and school absenteeism in children, jeopardizing future generations' economic development (Chen et al. 2018; Sunyer et al. 2015; Calderón-Garcidueñas et al. 2011; Currie et al. 2009). Furthermore, there is evidence that air pollution increases delinquent behavior, particularly among teenagers whose brains are still developing (Bondy et al. 2020; Burkhardt et al. 2019; Lu et al. 2018). Overall, air pollution has an impact on people's well-being. Life satisfaction decreases. People's beliefs, attitudes, judgments, and feelings change, causing them to become stressed, irritated, and upset, resulting in less emotional well-being and hedonic happiness, which fuels social unrest (Li et al. 2019; Zhang et al. 2017; Orru et al. 2016; Ferreira et al. 2013). Figure 1.1 depicts the effects of air pollution on public health and social well-being.

There is no doubt that air pollution is costly and undermines people's quality of life, but it has unfortunately been neglected in many countries around the world, particularly in low-income and middle-income countries. Industrial emissions, vehicular exhaust, and smoke-haze from slash-and-burn agriculture, as well as waste open

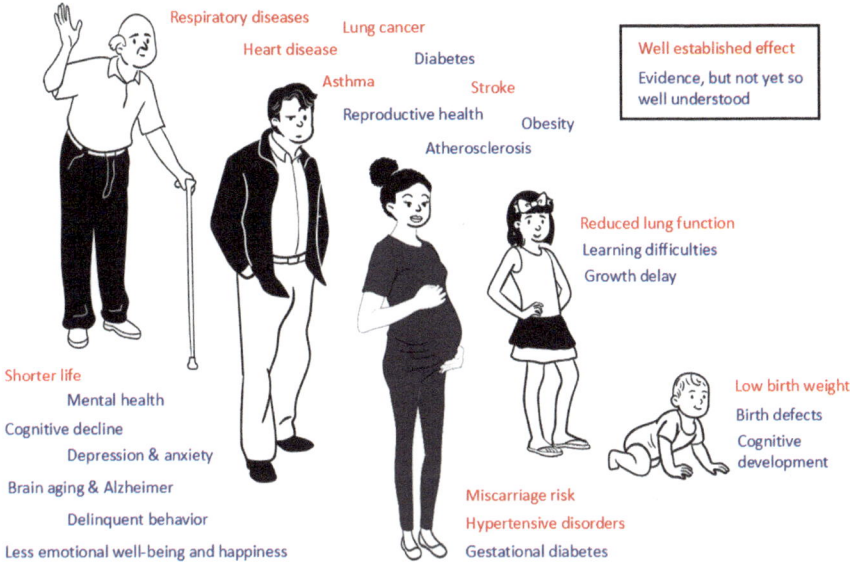

Fig. 1.1 Health effects linked to air pollution

burning, have especially been overlooked, despite their impact on public health, the economy, and the environment, leaving alone its close connection with climate change. Fuel combustion is responsible for the majority of particle pollution, as well as pollution from sulfur and nitrogen oxides, but it is also a major source of greenhouse gases (GHG) and short-lived climate pollutants that contribute to climate change. The Lancet Commission on Pollution and Health provides a comprehensive overview of the threat posed by air pollution to contemporary societies (Landrigan et al. 2018).

Air pollution is worsening in Southeast Asia, as it is in many other parts of the world. Rapid population growth, hasty industrialization, uncontrolled urban expansion, increased energy consumption and motorization, aggressive deforestation, and expansive agriculture, along with a changing climate and an underdeveloped governance system, are the culprits for air pollution alarmingly exceeding the daily and annual threshold levels recommended for health protection in large cities and many rural areas throughout the region.

The cost of air pollution in Southeast Asia can no longer be ignored. Governments must make pollution control a top priority on their agendas. As a result, there is a greater need for accurate information about air quality. Nonetheless, there are significant gaps in information and knowledge about the state of air quality in the region. Despite growing recognition of the threat posed by air pollution, no current efforts are underway to characterize the problem and determine whether the environmental policies implemented thus far are on track to improve air quality.

The COVID-19 pandemic served as a reminder of the value of breathing clean air. The health effects were more severe for those who contracted the virus in polluted areas, increasing their risk of hospitalization, length of stay, and likelihood of passing away from the disease (Vos et al. 2023; Zhang et al. 2023; Weaver et al. 2022). Southeast Asia, like many other parts of the world, saw the underlying toll that air pollution took on those suffering from respiratory and cardiovascular symptoms, thus increasing the vulnerability of society to the effects of the virus (Pozzer et al. 2020). We also learned that the radical approach of shutting down productive sectors associated with the emission of air pollutants, as was done as part of the social and economic restrictions imposed to curb the virus' spread, is not an efficient sustainable solution to achieve clean air. The restrictions temporarily improved air quality in terms of pollutant species of primary origin (i.e., those that are directly released to the atmosphere by emission sources), such as carbon monoxide (CO), sulfur dioxide (SO_2), nitrogen dioxide (NO_2), and the fraction of large particles ($\leq 10\ \mu$m in size, PM_{10}). However, pollutant species of secondary origin (i.e., those formed in the atmosphere through chemical reactions), such as ozone (O_3) and a fraction of fine particles ($\leq 2.5\ \mu$m in size, $PM_{2.5}$) did not exhibit changes, but in specific cases, well-defined increases were observed in response to change in the mix of precursors species (Sokhi et al. 2021).

Improving air quality requires addressing measurement, data, and reporting issues. To achieve clean air, it is necessary to collect reliable data on the abundance, origin, transformation, dispersion, and fate of the various pollutant species. Without reliable and accurate data, it is difficult to determine the air quality conditions, assess the effectiveness of any emission control measures, and understand unexpected outcomes like those observed during the COVID-19 lockdowns.

Along with implementing measures to improve air quality, governments must develop tools to evaluate their progress. These are the tools that support the air quality management process, and they include the application of formal risk assessments that take into account atmospheric processes, ambient measurements, emissions characterization, air quality modeling of emissions to ambient concentrations, and characterization of human and ecological responses to pollutants exposure (Hidy et al. 2011). According to the United States-Environmental Protection Agency (US-EPA), air quality management refers to all of the activities that a regulatory authority undertakes to help protect human health and the environment from the harmful effects of air pollution. As illustrated in Fig. 1.2, the air quality management process can be viewed as a cycle of interrelated actions supported by scientific research and technical assignments.

The air quality management process typically begins after authorities have set air quality goals and targets and established threshold concentrations for key pollutant species that will protect public health and the environment (i.e., air quality standards). Then, using the assessment tools listed above, they must determine the emission reductions required to meet the air quality targets and standards. Air quality managers should include compliance and implementation plans with a timeline when developing control strategies. To successfully achieve the required reductions, authorities will need to implement programs and enforce rules and regulations, as well as conduct ongoing evaluations to assess the effectiveness of the strategies and track progress toward meeting the air quality goals. To take informed decisions, air quality

Fig. 1.2 The air quality management process cycle. Goals and strategies must be reviewed on a regular basis, making the entire cycle a dynamic process. Each component of the cycle must be supported by scientific research so that air quality managers can take informed action based on a clear understanding of the origin, transformation, dispersion, and fate of air pollutants, as well as their impacts on human health and the ecosystem

managers must also use scientific information derived from ambient monitoring, data analysis, field and laboratory studies, and the development of new methodologies.

This monograph examines and evaluates the state of scientific research and air quality management in Southeast Asia within this context. It aims to serve as a starting point for a science-based conversation about implementing solutions to Southeast Asia's air pollution problem by bringing together an analysis of the availability of air quality data and studies from across the region.

The monograph is written with the broad range of actors in the air quality management process in mind, including policymakers, air quality managers, practitioners, the scientific community, and the non-governmental community. The information should also be useful for teachers and students, as well as citizens looking for information about air quality in the region.

Instead of serving as a reference that intrinsically describes the state of air quality, the monograph has been prepared as a reference that enumerates and critically examines the sources of data on air quality in Southeast Asia. Nevertheless, Chapter 3 provides a brief overview of the region's overall air quality.

Similarly, the monograph examines scientific work on the development of regional and local capacity for air quality monitoring, the construction of emission inventories, and the use of atmospheric chemical-transport models (air quality models in short) for operational, research, and forecasting purposes, as well as the use of remote sensing and satellite observations to address air pollution at various spatial and temporal scales. The work also tracks the scientific literature on studies about other issues related to air quality in Southeast Asia, such as studies on the physicochemical characterization of particles and other pollutant species, especially those resulting from biomass burning, as well as studies on the origin, seasonal patterns, trends, origin, and impact on public health of atmospheric pollution.

As a preamble to understanding the challenges facing air quality management, the work begins by briefly presenting the geographical and climatological conditions, as well as the economic and social situation of Southeast Asia in Chapter 2. Chapter 3 summarizes the current state of air quality in the region by drawing on key international assessments such as the State of Global Air 2024 (HEI 2024), the Global Burden of Diseases 2024 (IHME 2024), the Global Health Cost of Air Pollution (World Bank 2022), and the Emissions Database for Global Atmospheric Research (EDGAR, Crippa et al. 2020, 2024), as well as scientific literature.

After a brief explanation of the methodology and data sources used for the aforementioned international assessments, as well as their potential uncertainties for the case of Southeast Asia in Chapter 4, the focus of the work then shifts to the specifics of air quality management in each country in the region. The available data derived from the infrastructure currently in place to support air quality management is examined. The status of air quality monitoring in each country is reviewed in Chapter 5, including air quality monitoring networks operated for regulatory and surveillance purposes, and the deployment of low-cost sensors to complement existing air quality monitoring networks, as well as ambient air quality networks used for other purposes. The emission inventories compiled by authorities are then reviewed in Chapter 6, as are the gridded emission inventories built by academia and international consortia focused on the construction of global and regional emission inventories. Chapter 7 reviews the use of air quality models in research projects and international initiatives to forecast air pollution at global and regional scales. The use of satellite instrumentation to monitor air quality, including its challenges, is discussed in Chapter 8. Chapter 9 reviews the scientific initiatives taken over the last decade to better understand the region's air pollution problem.

Chapter 10 summarizes the challenges facing Southeast Asia's air quality management and discusses how the region's eleven nations can cooperate to address technical deficiencies and close data gaps. Based on the findings of the research conducted for this work, lessons learned from other parts of the world, and the authors' experience in air quality management, this chapter offers a set of recommendations that are specifically tailored to the needs and conditions of Southeast Asia. Chapter 11 closes the monograph with a series of recommendations on how policymakers, air quality managers, scientists, educators, professional communicators and journalists, and the non-governmental community and society in general could collaborate to achieve clean air with a local and regional vision.

References

Apte, J.S., Brauer, M., Cohen, A.J., Ezzati, M., Pope III, C.A.: Ambient $PM_{2.5}$ reduces global and regional life expectancy. Environ. Sci. Technol. Lett. **5**(9), 546–551 (2018). https://doi.org/10.1021/acs.estlett.8b00360

Atkinson, R.W., Anderson, H.R., Sunyer, J., Ayres, J., Baccini, M., Vonk, J.M., Boumghar, A., Forastiere, F., Forsberg, B., Touloumi, G., Schwartz, J., Katsouyanin, K.: Acute effects of particulate air pollution on respiratory admissions: results from APHEA 2 project. Am. J. Respir. Crit. Care Med. Respir. Crit. Care Med. **164**(10), 1860–1866 (2001). https://doi.org/10.1164/ajrccm.164.10.2010138

Bondy, M., Roth, S., Sager, L.: Crime is in the air: the contemporaneous relationship between air pollution and crime. J. Assoc. Environ. Resour. Econ. **7**(3), 417–618 (2020). https://doi.org/10.1086/707127

Boogaard, H., Walker, K., Cohen, A.J.: Air pollution: the emergence of a major global health risk factor. Int. Health **11**(6), 417–421 (2019). https://doi.org/10.1093/inthealth/ihz078

Brook, R.D., Rajagopalan, S., Pope, C.A., III., Brook, J.R., Bhatnagar, A., Diez-Roux, A.V., Holguin, F., Hong, Y., Luepker, R.V., Mittleman, M.A., Peters, A., Siscovick, D., Smith, S.C., Jr., Whitsel, J., Kaufman, J.D.: Particulate matter air pollution and cardiovascular disease: an update to the scientific statement from the American heart association. Circulation **121**(21), 2331–2378 (2010). https://doi.org/10.1161/CIR.0b013e3181dbece1

Brunekreef, B., Holgate, S.T.: Air pollution and health. Lancet **360**(9341), 1233–1242 (2002). https://doi.org/10.1016/S0140-6736(02)11274-8

Burkhardt, J., Bayham, J., Wilson, A., Carter, E., Berman, J.D., O'Dell, K., Ford, B., Fischer, E.V., Pierce, J.R.: The effect of pollution on crime: Evidence from data on particulate matter and ozone. J. Environ. Econ. Manag. **98**, 102267 (2019). https://doi.org/10.1016/j.jeem.2019.102267

Calderón-Garcidueñas, L., Calderón-Garcidueñas, A., Torres-Jardón, R., Avila-Ramírez, J., Kulesza, R.J., Angiulli, A.D.: Air pollution and your brain: what do you need to know right now. Primary Health Care Res. Dev. **16**(4), 329–345 (2015). https://doi.org/10.1017/S1463423614000036X

Calderón-Garcidueñas, L., Engle, R., Mora-Tiscareño, A., Styner, M., Gómez-Garza, G., Zhu, H., Jewells, V., Torres-Jardón, R., Romero, L., Monroy-Acosta, M.E., Bryant, C., González-González, Medina-Cortina, H., D'Angiulli, A.: Exposure to severe urban air pollution influences cognitive outcomes, brain volume and systemic inflammation in clinically healthy children. Brain Cogn. **77**(3), 345–355 (2011). https://doi.org/10.1016/j.bandc.2011.09.006

Capello, F., Gaddi, A.V.: Clinical Handbook of Air Pollution-Related Diseases. Springer, Switzerland. ISBN 978-3-319-62730-4 (2018). https://doi.org/10.1007/978-3-319-62731-1

Capello, F., Pili, G.: Air pollution in infancy, childhood and young adults. In: Capello, F., Gaddi, A. (eds.). Clinical Handbook of Air Pollution-Related Diseases. Springer, Cham (2018). https://doi.org/10.1007/978-3-319-62731-1_10

Chen, F., Chen, Z.: Cost of economic growth: air pollution and health expenditure. Sci. Total. Environ. **755**, 142543 (2021). https://doi.org/10.1016/j.scitotenv.2020.142543

Chen, S., Guo, C., Huang, X.: Air pollution, student health, and school absences: evidence from China. J. Environ. Econ. Manag. **92**, 465–497 (2018). https://doi.org/10.1016/j.jeem.2019.102267

Crippa, M., Solazzo, E., Huang, G., Guizzardi, D., Koffi, E., Muntean, M., Schieberle, C., Friedrich, R., Janssens-Maenhout, G.: High resolution temporal profiles in the emissions database for global atmospheric research. Sci. Data **7**, 121 (2020). https://doi.org/10.1038/s41597-020-0462-2

Crippa, M., Guizzardi, D., Pagani, F., Schiavina, M., Melchiorri, M., Pisoni, E., Graziosi, F., Muntean, M., Maes, J., Dijkstra, L., Van Damme, M., Clarisse, L., and Coheur, P.: Insights on the spatial distribution of global, national and sub-national GHG emissions in EDGARv8.0. Earth Syst. Sci. Data **16**(6), 2811–2830 (2024). https://doi.org/10.5194/essd-16-2811-2024

Currie, J., Hanushek, E.A., Kahn, E.M., Neidell, M., Rivkin, S.G.: Dos pollution increase school absences? Rev. Econ. Stat. **91**(4), 682–694 (2009). https://doi.org/10.1162/rest.91.4.682

Fardel, O., Kolasa, E., Le Vee, M.: Environmental chemicals as substrates, inhibitors or inducers of drug transporters: implication for toxicokinetics, toxicity and pharmacokinetics. Expert Opin. Drug Metab. Toxicol. **8**(1), 29–46 (2012). https://doi.org/10.1517/17425255.2012.637918

Ferreira, S., Akay, A., Brereton, F., Cuñado, J., Martinsson, P., Moro, M., Ningal, T.F.: Life satisfaction and air quality in Europe. Ecol. Econ. **88**, 1–10 (2013). https://doi.org/10.1016/j.ecolecon.2012.12.027

Franklin, B.A., Brook, R., Pope, A.C. III.: Air pollution and cardiovascular disease. Current Probl. Cardiol. **40**(5), 207–238 (2015). https://doi.org/10.1016/j.cpcardiol.2015.01.003

Goldizen, F.C., Sly, P.D., Knibbs, L.D.: Respiratory effects of air pollution on children. Pediatr. Pulmonol. **51**(1), 94–108 (2016). https://doi.org/10.1002/ppul.23262

Graff Zivin, J., Neidell, M.: The impact of pollution on worker productivity. Am. Econ. Rev. **102**(7), 3652–3673 (2012). https://doi.org/10.1257/aer.102.7.3652

Ha, S., Sundaram, R., Louis, G.M.B., Nobles, C., Seeni, I., Sherman, S., Mendola, P.: Ambient air pollution and the risk of pregnancy loss: a prospective cohort study. Fertil. Steril. **109**(1), 148–153 (2018). https://doi.org/10.1016/j.fertnstert.2017.09.037

Ha, S., Yeung, E., Bell, E., Insaf, T., Ghassabian, A., Bell, G., Muscatiello, N., Mendola, P.: Prenatal and early life exposures to ambient air pollution and development. Environ. Res. **174**, 170–175 (2019). https://doi.org/10.1016/j.envres.2019.03.064

Hanna, R., Oliva, P.: The effect of pollution on labor supply: Evidence from a natural experiment in Mexico city. J. Public Econ. **122**, 68–79 (2015). https://doi.org/10.1016/j.jpubeco.2014.10.004

Health Effects Institute (HEI).: State of Global Air 2024. Special Report. Health Effects Institute, Boston, MA (2024). ISSN 2578–6873. https://www.stateofglobalair.org/

Hidy, G.M., Brook, J.R., Demerjian, K.L., Molina, L.T., Pennell, W.T., Scheffe, R.D.: Technical challenges of multipollutant air quality management, Springer, Dordrecht, The Netherlands (2011). ISBN: 978-94-007-0304-9. https://doi.org/10.1007/978-94-007-0304-9

Hoek, G., Krishnan, R.M., Beelen, R., Peters, A., Ostro, B., Brunekreef, B., Kaufman, J.D.: Long-term air pollution exposure and cardio-respiratory mortality: a review. Environ. Health **12**, 43 (2013). https://doi.org/10.1186/1476-069X-12-43

Hu, C.Y., Gao, X., Fang, Y., Jiang, W., Huang, K., Hua, X.G., Yang, X.J., Chen, H.B., Jiang, Z.X., Zhang, X.J.: Human epidemiological evidence about the association between air pollution exposure and gestational diabetes mellitus: systematic review and meta-analysis. Environ. Res. **180**, 108843 (2020). https://doi.org/10.1016/j.envres.2019.108843

Institute for Health Metrics and Evaluation (IHME).: Global Burden of Diseases Compare Data Visualization. Seattle, WA. IHME, University of Washington (2024). http://vizhub.healthdata.org/gbd-compare

Johnson, N.M., Hoffmann, A.R., Behlen, J.C., Lau, C., Pendleton, D., Harvey, N., Shore, R., Li, Y., Chen, J., Tian, Y., Zhang, R.: Air pollution and children's health—a review of adverse effects associated with prenatal exposure from fine to ultrafine particulate matter. Environ. Health Prev. Med. **26**, 72 (2021). https://doi.org/10.1186/s12199-021-00995-5

La Marca, L., Gava, G.: Air pollution effects in pregnancy. In: Capello, F., Gaddi, A. (eds.). Clinical Handbook of Air Pollution-Related Diseases. Springer, Cham (2018). https://doi.org/10.1007/978-3-319-62731-1_26

Landrigan, P.J., Fuller, R., Acosta, N.J., Adeyi, O., Arnold, R., Baldé, A.B., Bertollini, R., Bose-O'Reilly, S., Boufford, J.I., Breysse, P.N., Chiles, T., et al.: The Lancet commission on pollution and health. Lancet **391**(10119), 462–512 (2018). https://doi.org/10.1016/S0140-6736(17)32345-0

Landrigan, P.J., Fuller, R., Fisher, S., Suk, W.A., Sly, P., Chiles, T.C., Bose-O'Reilly, S.: Pollution and children's health. Sci. Total. Environ. **650**, 2389–2394 (2019). https://doi.org/10.1016/j.scitotenv.2018.09.375

Le Vee, M., Jouan, E., Stieger, B., Lecureur, V., Fardel, O.: Regulation of human hepatic drug transporter activity and expression by diesel exhaust particle extract. PLoS ONE **10**(3), e0121232 (2015). https://doi.org/10.1371/journal.pone.0121232

Lertxundi, A., Baccini, M., Lertxundi, N., Fano, E., Aranbarri, A., Martínez, M.D., Ayerdi, M., Álvarez, J., Santa-Marina, L., Dorronsoro, M., Ibarluzea, J.: Exposure to fine particle matter, nitrogen dioxide and benzene during pregnancy and cognitive and psychomotor developments in children at 15 months of age. Environ. Int. **80**, 33–40 (2015). https://doi.org/10.1016/j.envint.2015.03.007

Li, Y., Guan, D., Yu, Y., Westland, S., Wang, D., Meng, J., Wang, X., He, K., Tao, S.: A psychophysical measurement on subjective well-being and air pollution. Nat. Commun. **10**, 1–8 (2019). https://doi.org/10.1038/s41467-019-13459-w

Link, M.S., Luttmann-Gibson, H., Schwartz, J., Mittleman, M.A., Wessler, B., Gold, D.R., Dockery, D.W., Laden, F.: Acute exposure to air pollution triggers atrial fibrillation. J. Am. Coll. Cardiol. **62**(9), 816–825 (2013). https://doi.org/10.1016/j.jacc.2013.05.043

Liu, Y., Pan, J., Fan, C., Xu, R., Wang, Y., Xu, C., Xie, S., Zhang, H., Cui, X., Peng, Z., Shi, C., Zhang, Y., Sun, H., Zhou, Y., Zhang, L.: Short-term exposure to ambient air pollution and mortality from myocardial infarction. J. Am. Coll. Cardiol. **77**(3), 271–281 (2021). https://doi.org/10.1016/j.jacc.2020.11.033

Lu, J.G., Lee, J.J., Gino, F., Galinsky, A.D.: Polluted morality: air pollution predicts criminal activity and unethical behavior. Psychol. Sci. **29**(3), 340–355 (2018). https://doi.org/10.1177/0956797617177358

Lu, W., Tian, Q., Xu, R., Zhong, C., Qiu, L., Zhang, H., Shi, C., Liu, Y., Zhou, Y.: Short-term exposure to ambient air pollution and pneumonia hospital admission among patients with COPD: a time-stratified case-crossover study. Respir. Res. **23**, 71 (2022). https://doi.org/10.1186/s12931-022-01989-9

McGrath, L., Hynes, S., McHale, J.: The air we breathe: estimates of air pollution extended genuine savings for Europe. Rev. Income Wealth **68**(1), 161–188 (2022). https://doi.org/10.1111/roiw.12512

Orru, K., Orru, H., Maasikmets, M., Hendrikson, R., Ainsaar, M.: Well-being and environmental quality: Does pollution affect life satisfaction? Qual. Life Res. **25**, 699–705 (2016). https://doi.org/10.1007/s11136-015-1104-6

Paul, L.A., Burnett, R.T., Kwong, J.C., Hystad, P., van Donkelaar, A., Bai, L., Goldberg, M.S., Lavigne, E., Copes, R., Martin, R.V., Kopp, A.: The impact of air pollution on the incidence of diabetes and survival among prevalent diabetes cases. Environ. Int. **134**, 105333 (2020). https://doi.org/10.1016/j.envint.2019.105333

Payne-Sturges, D.C., Marty, M.A., Perera, F., Miller, M.D., Swanson, M., Ellickson, K., Cory-Slechta, D.A., Ritz, B., Balmes, J., Anderko, L., Talbott, E.O.: Healthy air, healthy brains: advancing air pollution policy to protect children's health. Am. J. Public Health **109**(4), 550–554 (2019). https://doi.org/10.2105/AJPH.2018.304902

Pozzer, A., Dominici, F., Haines, A., Witt, C., Münzel, T., Lelieveld, J.: Regional and global contributions of air pollution to risk of death from COVID-19. Cardiovasc. Res. **116**(14), 2247–2253 (2020). https://doi.org/10.1093/cvr/cvaa288

Pun, V.C., Manjourides, J., Suh, H.: Association of ambient air pollution with depressive and anxiety symptoms in older adults: results from the NSHAP study. Environ. Health Perspect. **125**(3), 342–348 (2017). https://doi.org/10.1289/EHP494

Ronaldson, A., Arias de La Torre, J., Ashworth, M., Hansell, A.L., Hotopf, M., Mudway, I., Stewart, R., Dregan, A., Bakolis, I.: Associations between air pollution and multimorbidity in the UK Biobank: a cross-sectional study. Front. Public Health **10**, 1035415 (2022). https://doi.org/10.3389/fpubh.2022.1035415

Sabatini, D.: Air pollution and elderly. In: Capello, F., Gaddi, A. (eds) Clinical Handbook of Air Pollution-Related Diseases. Springer, Cham (2018). https://doi.org/10.1007/978-3-319-62731-1_11

Shah, A.S., Lee, K.K., McAllister, D.A., Hunter, A., Nair, H., Whiteley, W., Langrish, J.P., Newby, D.E., Mills, N.L.: Short term exposure to air pollution and stroke: systematic review and meta-analysis. BMJ **350**, h1295 (2015). https://doi.org/10.1136/bmj.h1295

Shi, L., Zhu, Q., Wang, Y., Hao, H., Zhang, H., Schwartz, J., Amini, H., van Donkelaar, A., Martin, R.V., Steenland, K., Sarnat, J.A., Caudel, W.M., Ma, T., Li, H., Chang, H.H., Wingo, T., Mao, X., Russell, A.G., Weber, R.J., Liu, P.: Incident dementia and long-term exposure to constituents of fine particle air pollution: a national cohort study in the United States. Proc. Natl. Acad. Sci. **120**(1), 2211282119 (2022). https://doi.org/10.1073/pnas.2211282119

Simoni, M., Baldacci, S., Maio, S., Cerrai, S., Sarno, G., Viegi, G.: Adverse effects of outdoor pollution in the elderly. J. Thorac. Dis. **7**(1), 34–45 (2015). https://doi.org/10.3978/j.issn.2072-1439.2014.12.10

Sinharay, R., Gong, J., Ohman-Strickland, P., Ernst, S., Kelly, F.J., Zhang, J., Collins, P., Cullinan, P., Chung, K.F.: Respiratory and cardiovascular responses to walking down a traffic-polluted road compared with walking in a traffic-free area in participants aged 60 years and older with chronic lung or heart disease and age matched healthy controls: a randomized, crossover study. Lancet **391**(10118), 339–349 (2018). https://doi.org/10.1016/S0140-6736(17)32643-0

Sokhi, R.S., Singh, V., Querol, X., Finardi, S., Targino, A.C., de Fatima Andrade, M., Pavlovic, R., Garland, R.M., Massagué, J., Kong, S., Baklanov, A., et al.: A global observational analysis to understand changes in air quality during exceptionally low anthropogenic emission conditions. Environ. Int. **157**, 106818 (2021). https://doi.org/10.1016/j.envint.2021.106818

Sunyer, J., Esnaola, M., Alvarez-Pedrerol, M., Forns, J., Rivas, I., López-Vicente, M., Suades-González, E., Foraster, M., Garcia-Esteban, R., Basagaña, X., Viana, M., Cirach, M., Moreno, T., Alastuey, A., Sebastian-Galles, N., Nieuwenhuijsen, M., Querol, X.: Association between traffic-related air pollution in schools and cognitive development in primary school children: a prospective cohort study. PLoS Med. **12**(3), e1001792 (2015). https://doi.org/10.1371/journal.pmed.1001792

Tumiatti, V., Fimognari, C., Milelli, A., Manucra, D.: Pollutants and drugs: Interactions and human health. In: Capello, F., Gaddi, A. (eds.) Clinical Handbook of Air Pollution-Related Diseases. Springer, Cham (2018). https://doi.org/10.1007/978-3-319-62731-1_12

Turner, M.C., Andersen, Z.J., Baccarelli, A., Diver, W.R., Gapstur, S.M., Pope III, C.A., Prada, D., Samet, J., Thurston, G., Cohen, A.: Outdoor air pollution and cancer: an overview of the current evidence and public health recommendations. CA: Cancer J. Clin. **70**(6), 460–479 (2020). https://doi.org/10.3322/caac.21632

Vos, S., De Waele, E., Goeminne, P., Bijnens, E.M., Bongaerts, E., Martens, D.S., Malina, R., Ameloot, M., Dams, K., De Weerdt, A., Dewyspelaere, G., Jacobs, R., Mistiaen, G., Jorens, P., Nawrot, T.S.: Pre-admission ambient air pollution and blood soot particles predict hospitalisation outcomes in COVID-19 patients. Eur. Respir. J.Respir. J. **63**, 2300309 (2023). https://doi.org/10.1183/13993003.00309-2023

Weaver, A.K., Head, J.R., Gould, C.F., Carlton, E.J., Remais, J.V.: Environmental factors influencing COVID-19 incidence and severity. Annu. Rev. Public Health **43**, 271–291 (2022). https://doi.org/10.1146/annurev-publhealth-052120-101420

Wellenius, G.A., Burger, M.R., Coull, B.A., Schwartz, J., Suh, H.H., Koutrakis, P., Schlaug, G., Gold, D.R., Mittleman, M.A.: Ambient air pollution and the risk of acute ischemic stroke. Arch. Intern. Med. **172**(3), 229–234 (2012). https://doi.org/10.1001/archinternmed.2011.732

World Bank (2022). The global health cost of $PM_{2.5}$ air pollution: a case for action beyond 2021. In: International Development in Focus. World Bank, Washington, DC (2022)

Wu, J., Laurent, O., Li, L., Hu, J., Kleeman, M.: Adverse reproductive health outcomes and exposure to gaseous and particulate matter air pollution in pregnant women. Research Report 188. Health Effects Institute. Boston MA, USA (2016). https://www.healtheffects.org/system/files/Wu-RR188.pdf

Yap, J., Ng, Y., Yeo, K.K., Sahlén, A., Lam, C.S.P., Lee, V., Ma, S.: Particulate air pollution on cardiovascular mortality in the tropics: impact on the elderly. Environ. Health **18**, 34 (2019). https://doi.org/10.1186/s12940-019-0476-4

Zhang, J., Lim, Y.H., So, R., Jørgensen, J.T., Mortensen, L.H., Napolitano, G.M., Cole-Hunter, T., Loft, S., Bhatt, S., Hoek, G., Brunekreef, B., Westendorp, R., Ketzel, M., Jørgen, B., Lange, T., Kølsen-Fisher, T., Andersen, J.Z.: Long-term exposure to air pollution and risk of SARS-CoV-2 infection and COVID-19 hospitalization or death: Danish nationwide cohort study. Eur. Respir. J. **62**, 2300280 (2023). https://doi.org/10.1183/13993003.00280-2023

Zhang, X., Zhang, X., Chen, X.: Happiness in the air: How does a dirty sky affect mental health and subjective well-being? J. Environ. Econ. Manag. **85**, 81–94 (2017). https://doi.org/10.1016/j.jeem.2017.04.001

Chapter 2
Southeast Asia

Abstract This chapter provides a brief overview of Southeast Asia and introduces the Association of Southeast Asian Nations (ASEAN), assisting readers to better understand the region's air quality challenges. It reviews the geography and climate of the region and looks into the economic development, industrialization, and urbanization, as well as the resulting increase in energy consumption. It also presents some of the governance and social challenges that the eleven ASEAN member countries face in their highly diverse and unequal societies.

Keywords Association of Southeast Asian Nations · Economic development · Industrialization · Urbanization · Governance · Southeast Asia society

Southeast Asia is the geographical southeastern region of Asia (see Fig. 2.1), covering approximately 4.5-million km^2 (3% of Earth's total land area), housing nearly 675 million people (8.5% of world's population), and consisting of eleven countries divided into two subregions:

- Mainland Southeast Asia, also known as Indochina, comprises Cambodia, Lao PDR, Myanmar, Peninsular Malaysia, Thailand, and Vietnam.
- Maritime Southeast Asia, mostly consisting of the Malay Archipelago or *Nusantara*, comprises Brunei Darussalam, Indonesia, the island portion of Malaysia, Singapore, the Philippines, and Timor-Leste.

The Association of Southeast Asian Nations (ASEAN) is made up of eleven countries, with Timor-Leste currently serving as an observer. ASEAN is a regional organization established to promote intergovernmental cooperation and facilitate economic, political, security, educational, and cultural integration among its members. Its main objective is to accelerate economic growth and, through it, social progress, and sociocultural development. It also aims to promote regional peace and stability through abiding respect for justice and the rule of law in the relationship among countries in the region and adherence to the principles of the United Nations Charter. Each and every ASEAN member is dedicated to achieving the 2030 Agenda for

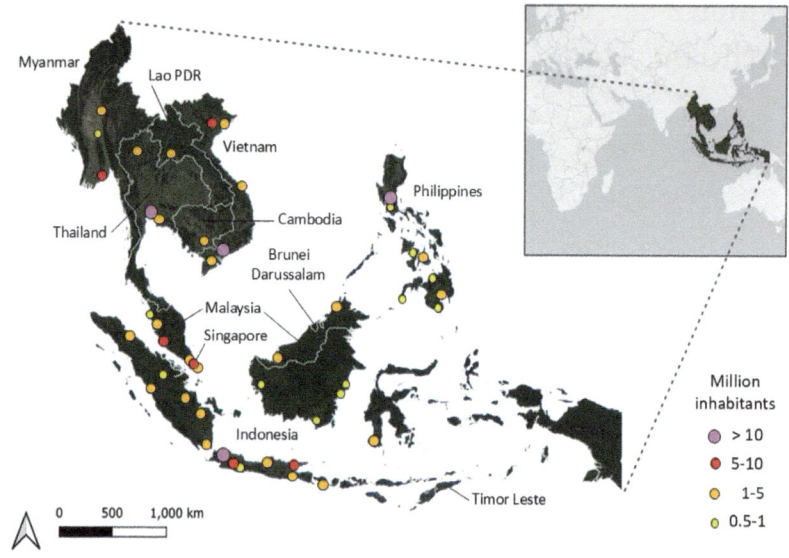

Fig. 2.1 Map of Southeast Asia pointing out cities with more than 500,000 residents

Sustainable Development in terms of environmental protection (https://sdgs.un.org/2030agenda), including the Sustainable Development Goal Indicator 11.6.2 referring to the annual mean concentration of fine particles ($PM_{2.5}$) in cities.

2.1   Economy

The gross domestic product (GDP) of the eleven countries that make up Southeast Asia totaled US$3.35 trillion in 2021, or 3.9% of the global GDP. Southeast Asia is the world's fifth-largest economy, trailing only the United States, China, Japan, and Germany. In 2019, before the COVID-19 pandemic, the region's annual average GDP growth was 5.3%, outpacing the global average of 2.6%, and taken as a whole the region is on track to become the world's fourth-largest economy by 2030. As a result, poverty has decreased, with the percentage of the population living below the national poverty line falling from 26.8% in 2005 to 6.2% in 2019 (ASEAN 2021a). However, as evidenced by the Gini Ratio, there is still a need to address income inequality (0 means perfect equality and 1 means perfect inequality in income distribution). Only Cambodia has seen a significant reduction in inequality, with its Gini Ratio falling from 0.42 in 2005 to 0.29 in 2019. Malaysia, the Philippines, Singapore, Thailand, and Vietnam have also reduced their Gini Ratio, but by less than 0.05, remaining in the 0.40–0.45 range. In contrast, inequality in Indonesia and Lao PDR

Table 2.1 Selected information and key statistics for each country in Southeast Asia

	Brunei Darussalam	Cambodia	Indonesia	Laos PDR	Malaysia	Myanmar	Philippines	Singapore	Thailand	Timor-Leste	Vietnam
Total area (km²)[a]	5765	181,035	1,904,569	236,800	329,847	676,578	300,000	719	513,120	14,874	331,210
Population, 2021[a]	441,530	16,946,450	276,361,790	7,379,360	32,776,190	54,806,010	111,046,910	5,453,570	69,950,840	1,343,880	98,168,830
Population density (inhabitants per km²)[a]	77	94	145	31	99	81	370	7,585	136	90	296
GDP nominal, 2021 (million US dollars)[b]	14,007	26,961	1,186,093	18,827	372,701	65,068	394,086	396,987	505,982	1,777	362,638
GDP per capita, 2021 (US dollars)[b]	31,723	1591	4292	2551	11,371	1187	3549	72,794	7,233	1,959	3,694
Total energy supply (2020)[b]	0.2	0.3	10.0	0.2	3.8	1.0	2.6	1.4	5.8	0.01	3.8
Gini ratio[a]	Not available	0.29	0.38	0.36	0.41	0.30	0.43	0.45	0.43	Not available	0.40
Fertility rate, 2020[a]	1.6	2.4	2.1	2.5	1.7	2.4	2.7	1.1	1.5	3.9	2.1
Under-five mortality rate per 1000 live births, 2020[a]	9.7	25.9	25.0	65.1	6.9	58.4	27.0	2.1	7.9	42.3	22.3

(continued)

Table 2.1 (continued)

	Brunei Darussalam	Cambodia	Indonesia	Laos PDR	Malaysia	Myanmar	Philippines	Singapore	Thailand	Timor-Leste	Vietnam
Life expectancy (years)[a]	77.4	72.0	71.5	67.0	74.9	66.6	72.7	83.9	75.7	69.5	73.7
Adult literacy rate, 2019[a]	97.3	81.9	96.0	70.4	95.0	88.9	96.3	97.1	93.8	68.1	96.7
Population with access to safe drinking water, 2020 (%)[a]	100	79.7	90.2	77.5	95.9	86.2	95.6	100	99.9	85.5	97.4
Population with access to improved sanitation, 2020 (%)[a]	93.0	80.4	79.5	75.3	99.7	80.1	81.1	100	98.7	57.8	94.0
Unemployment rate, 2020 (%)[b]	7.4	2.4	7.1	9.4	4.5	0.5	10.3	4.1	1.7	4.9	2.3
Official language[a]	Malay and English	Khmer	Bahasa Indonesia	Lao	Malay and English	Burmese	Tagalog and English	English, Malay, Mandarin, Tamil	Thai	Tetum and Portuguese	Vietnamese
Dominant climate(s)[d]	Af	Aw	Af and Am	Aw and Cwa	Af	Am, Aw, BSh and Cwa	Af and Am	Af	Af & Am	Aw	Am, Aw and Cwa

[a] ASEAN Key Figs. 2021 (ASEAN 2021c)
[b] The World Bank, https://data.worldbank.org
[c] Southeast Asia Energy Outlook (IEA 2022)
[d] Köppen-Geiger classification. Af: tropical rainforest; Am: tropical monsoon; Aw: tropical savannah; BSh: Arid, steppe, hot; Cwa: temperate, dry winter, hot summer

has increased slightly over the same time period (ASEAN 2021a). Table 2.1 shows selected demographic data and key statistics for each Southeast Asian country.

Agriculture is vital to the economy of Southeast Asia. Agriculture is practiced by more than two-thirds of the labor force in both Cambodia and Lao PDR. Rice and rubber have long been major export commodities, while oil palm has emerged in the last three to four decades. Southeast Asia's industrialization is a relatively recent phenomenon; much of the growth has only taken place since the early 1960s. Services and manufacturing are expanding. The region's largest economy, Indonesia, is developing into an emerging market. Rapid industrialization is taking place in Thailand, Malaysia, and the Philippines, which are rated as middle-income nations. Vietnam is following their steps, but agriculture still has a significant impact on the country's economy. Except for Singapore, where the manufacture of a variety of products, led by electric and electronic equipment, is dominant, agricultural processing is the most important in all ASEAN nations. Textiles and clothing are important in Thailand, Myanmar, and the Philippines, as is the chemical industry in Malaysia, Singapore, Thailand, and Indonesia. Tin production remains important in Malaysia, Thailand, and Indonesia, despite the discouraging effects of fluctuating market prices. The region has significant oil and natural gas reserves, particularly in Indonesia, Malaysia, and Brunei Darussalam. The eleven ASEAN member states are currently working together to build a single market and production base through a high-impact economy integration in the region (ASEAN 2021b). The free flow of goods, services, investments, capital, and labor is expected to enable the development of regional production networks and strengthen ASEAN's capacity as a global supply chain.

Trade is especially important in Southeast Asia. As of 2019, the combined value of merchandise trade (US$2.8 trillion) and services (US$778.6 billion) was roughly one-third that of the USA (ASEAN 2021c). Exports account for a small proportion of GDP in Cambodia, Myanmar, Vietnam, and Lao PDR, but a moderate proportion in Thailand, the Philippines, and Indonesia, while it is significant in Malaysia, Brunei Darussalam, and Singapore. Malaysia's main exports are petroleum products and oil palm. Food and manufactured goods account for nearly all Thailand's trade. Brunei Darussalam's economy is almost entirely based on petroleum exports. Because of its unique geographical location, economic policies, and highly educated labor force, Singapore's investments in the manufacturing and service sectors have greatly expanded and attracted multinational corporations.

2.2 Energy Consumption

Energy consumption in Southeast Asia has increased from 16 exajoule (1 EJ $= 1 \times 10^{18}$ J) in 2000 to 29 EJ in 2020, reflecting the region's rapid development over the last two decades (see Fig. 2.1a; IEA 2022). This increase has come at the expense of increases in the use of coal, oil, and natural gas, as well as renewable energy (Fig. 2.2b). Industry has seen the greatest economic growth, and thus the greatest increase in energy consumption of any sector. The cement and steel industries have been the primary drivers of this expansion. Since 2000, power generation has nearly

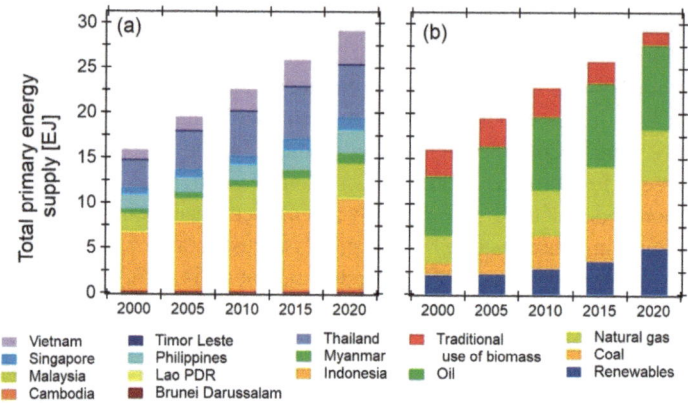

Fig. 2.2   Total primary energy supply by country (**a**) and by type of fuel (**b**) in Southeast Asia. Data obtained from Southeast Asia Energy Outlook 2022 (IEA 2022)

tripled, with coal-fired power plants accounting for the majority of the increase. In households, the traditional use of biomass has been displaced, mainly by electricity, as result of economic development and urbanization.

So far this century, the increase in the use of fossil fuels has accounted for more than 90% of total energy demand. Coal demand alone has expanded by a factor of six, and its share of total energy supply has increased from 8 to 26%. Oil demand has increased by more than 40%, despite the fact that its share of total energy supply has decreased from 42 to 32%. Natural gas consumption rose by more than 80% and has remained around 20% of the total energy mix. Today, the electricity and industrial sectors account for 70% of total natural gas consumption (IEA 2022).

Although the amount of energy produced by modern renewable sources of energy more than doubled between 2000 and 2020, their share of the total energy mix remains small, accounting for less than 10%. Despite the rapid growth of solar photovoltaic (PV) and wind systems in recent years, modern bioenergy, geothermal energy, and hydropower still account for more than 98% of total renewable energy in Southeast Asia today. Geothermal resources are primarily found in Indonesia and the Philippines. Cambodia, Lao PDR, and Myanmar have continued to develop domestic hydropower resources, utilizing their rivers and abundant rainfall.

Overall regional trends conceal some very different national-level situations in individual countries. The share of fossil fuels and reliance on coal varies greatly across the region; in manufacturing economies such as Thailand and Malaysia, the share of fossil fuels is higher, while it is lower in economies that continue to be more dependent on agriculture, like Myanmar and Lao PDR (see Fig. 2.2).

2.3 Governance

Most Southeast Asian countries have democratic forms of government, but good governance is still lacking in some of them. The region's confidence in government institutions has generally increased, but corruption, lack of transparency, and weak judicial oversight are still serious issues in some countries, as do ethnic tensions in divided societies, which argue for better public institutions to bridge gaps between groups.

Although South Asian countries have strong administrative capacities, the private sector handles many public functions and services. Furthermore, the business sector has a high concentration of ownership, which means that national policies are at the mercy of large corporations, which does not help reduce injustice and social inequality, nor does it contribute to building sustainable societies or protecting the environment. A welfare state does not yet exist in every country. Governments with better regulatory frameworks allocate their resources more efficiently, allowing them to make more investments and thus be better prepared to deal with financial uncertainties like those triggered by the COVID-19 pandemic, as Singapore did.

Unfortunately, this is not the case for most countries, which have poor fiscal management and rely on distorting trade taxes to fund public spending. Gonzalez and Mendoza (2002), Breslin and Nesadurai et al. (2018), and Khalid and Maidin (2022), among others, provide detailed perspectives on governance in Southeast Asia.

2.4 Society

The region is culturally and ethnically diverse. Nearly 1000 languages besides English are spoken across Southeast Asia; Indonesia alone has over 700 languages. People practice many different religions. No individual country is religiously homogeneous. By population, Islam is the most practiced faith followed by Buddhism and Christianity (ASEAN 2021c).

As a result of declining fertility and mortality rates, the region has experienced a decrease in annual population growth, from over 2.0% prior to 1992 to approximately 1.1% over the past five years. This phenomenon has altered the population age structure of Southeast Asia. Between 2000 and 2020, the youth population (0–19 years old) declined from 42.0% to 33.1% of the total population. On the contrary, the share of the productive working-age population (15–64 years old) has increased by 6%, reaching 60.0% nowadays. Likewise, the share of people aged 65 and up increased from 5.3% to 7.2%. Women currently have a life expectancy of 75.3 years, compared to 70.9 years for men (ASEAN 2021c).

In terms of literacy, six countries have achieved literacy rates that exceed 95% (Brunei, Singapore, Vietnam, the Philippines, and Malaysia). Similarly, more than 95% of children of primary school age do so, except in Cambodia (87%).

2.5   Urbanization

Southeast Asia's population has become increasingly urbanized, with the number of people living in cities increasing by approximately 70% since 2000. There are currently 35 cities with a population of over one million in the region (see Fig. 2.1). Jakarta is the most populous city, followed by Manila, Bangkok, Ho Chi Minh City, Kuala Lumpur, Hanoi, Bandung, Surabaya, Yangon, and Singapore. The first four cities are considered megacities because they have a population of more than 10-million inhabitants, the other six cities have a population of more than 5-million people. In Southeast Asia, almost 50% of people live in urban areas, which is lower than the global average of 55% (United Nations 2019).

A sharp rise in motorization has coincided with urbanization. Between 2000 and 2020, private vehicle ownership (passenger cars and motorbikes) more than quadrupled, rising from 27 to 59 vehicles per 1000 inhabitants. This has resulted in an 80% increase in oil demand in the region's transportation sector (IEA 2022).

Electricity consumption has increased as a result of urbanization. Buildings have seen the greatest increase in electricity consumption of any sector, accompanied by a significant increase in air conditioner and appliance use. For example, the number of air conditioning units has tripled in the last two decades, while the number of refrigerators has increased by 150% (IEA 2022).

2.6   Climate

Southeast Asia has a tropical climate that is hot and humid all year round, with plenty of rain. Northern Vietnam as well as mountainous parts of Lao PDR and Myanmar are the only regions with subtropical climate. The majority of Southeast Asia experiences wet and dry seasons as a result of seasonal monsoons.

Southeast Asia is one of the world's most vulnerable regions to climate change. Myanmar, the Philippines, and Thailand are among the countries that have experienced the highest number of fatalities and economic losses as a result of climate-related disasters (Ding and Beh 2022). Climate change will have a severe impact on agriculture since changes in rainfall and runoff will affect irrigation systems, affecting water quality and supply (Overland et al. 2017). However, efforts to reduce greenhouse gas emissions and build resilience to climate change disasters are sorely lagging (Overland et al. 2021).

References

Association of Southeast Asian Nations (ASEAN).: ASEAN Development outlook: inclusive and sustainable development. ASEAN secretariat. Jakarta, Indonesia. ISBN 978-602-5798-93-1 (2021a). https://asean.org

Association of Southeast Asian Nations (ASEAN).: Mid-term review: ASEAN economic community blueprint 2025. ASEAN secretariat, Jakarta, Indonesia. ISBN 978-623-6945-25-4 (2021b). https://asean.org

Association of Southeast Asian Nations (ASEAN).: ASEAN Key Figures 2021. ASEAN Secretariat, Jakarta, Indonesia. ISBN 978-623-6945-78-0 (2021c). https://asean.org

Breslin, S., Nesadurai, H.E.S.: Who governs and how? Non-state actors and transnational governance in Southeast Asia. J. Contemp. Asia **48**, 187–203 (2018). https://doi.org/10.1080/00472336.2017.1416423

Ding, D.K., Beh, S.E.: Climate change and sustainability in ASEAN countries. Sustainability **14**(2), 999 (2022). https://doi.org/10.3390/su14020999

Gonzalez, E.T., Mendoza, M.L.: Governance in Southeast Asia: issues and options (No. 2002–07). PIDS Discussion Paper Series (2002). https://www.jica.go.jp/jica-ri/IFIC_and_JBICI-Studies/jica-ri/publication/archives/jbic/report/paper/pdf/rp16_e07.pdf

International Energy Agency (IEA).: Southeast Asia Energy Outlook 2022. International Energy Agency Secretariat (2022). https://www.iea.org/reports/southeast-asia-energy-outlook-2022

Khalid, R.M., Maidin, A.J. (eds.): Good governance and the sustainable development goals in Southeast Asia. Routledge, Chennai, India. https://doi.org/10.4324/9781003230724

Overland, I. et al.: Impact of climate change on ASEAN international affairs: risk and opportunity multiplier, Norwegian Institute of International Affairs and Myanmar Institute of International and Strategic Studies (2017). https://www.researchgate.net/publication/320622312_Impact_of_Climate_Change_on_ASEAN_International_Affairs_Risk_and_Opportunity_Multiplier

Overland, I., Sagbakken, H.F., Chan, H.Y., Merdekawati, M., Suryadi, B., Utama, N.A., Vakulchuk, R.: The ASEAN climate and energy paradox. Energy Clim. Change **2**, 100019 (2021). https://doi.org/10.1016/j.egycc.2020.100019

United Nations.: Department of Economic and Social Affairs, Population Division. World Urbanization Prospects: the 2018 Revision. ST/ESA/SER.A/420. New York, United Nations (2019). https://population.un.org/wup/

Chapter 3
Air Pollution in Southeast Asia

Abstract Poor air quality is responsible for approximately 335,000 premature deaths in Southeast Asia each year, costing the region 6.75% of its GDP, according to estimates from the Health Effects Institute, the Institute for Health Metrics and Evaluation, and the World Bank. The burden of health damage and monetary cost is not distributed evenly across the region due to differences in economic development and social inequality, as discussed in this chapter in conjunction with trends of population-weighted ambient concentrations of O_3 and $PM_{2.5}$ at the country scale, and $PM_{2.5}$ and NO_2 at the city scale, as well as trends of pollutants emission as compiled in the Emissions Database for Global Atmospheric Research (EDGAR). Note that the estimates presented here are subject to uncertainty due to a lack of ground truth in many aspects, although they are likely the best source of information on the matter.

Keywords Air quality trends · Air pollutants emission · Premature deaths · Economic loss

Ambient (also called outdoor) air pollution is the culprit of about 335-thousand (95% UI: 192 to 487 thousand) premature deaths in Southeast Asia every year (see Fig. 3.1; HEI 2020). On average, 50 premature annual deaths per 100,000 inhabitants are attributed to air pollution. This figure ranges from 9 in Brunei Darussalam to 29 in Singapore, 49 in Indonesia, 64 in Myanmar, and 78 in Thailand.

In Southeast Asia, ambient air pollution contributed 6.2% to total mortality in 2021 and was the sixth highest risk factor for death, trailing only high blood pressure (20.1%), dietary risks (10.9%), smoking (9.9%), diabetes (8.4%) and kidney disease (7.1%).

The burden of diseases attributable to ambient air pollution is not evenly distributed throughout Southeast Asia. For example, while the death risk from chronic obstructive pulmonary disease (COPD) is greater than 25% in Thailand, it is only 5% in Brunei Darussalam. Similarly, the risk of diabetes death in Cambodia, Lao PDR, and Timor-Leste is 4–5 times lower than in Singapore and Thailand.

© The Author(s), under exclusive license to Springer Nature Switzerland AG 2024
E. Velasco et al., *Air Quality Management and Research in Southeast Asia*,
SpringerBriefs in Earth System Sciences, https://doi.org/10.1007/978-3-031-69088-4_3

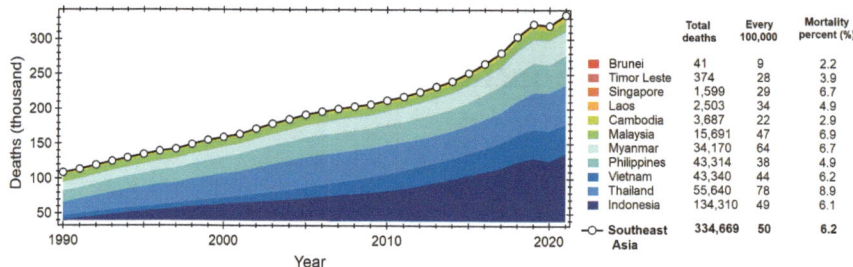

Fig. 3.1 Annual premature deaths likely attributed to PM$_{2.5}$ and O$_3$ pollution. The columns on the right list the total number of premature deaths and the number of premature deaths per 100,000 inhabitants for each country using data from the State of Global Air Report 2024 adjusted to 2021 (HEI 2024) and the percent of total deaths for the same year as computed by the Global Burden of Diseases 2024 (IHME 2024)

Country	COPD[a]	Stroke	Lung cancer	Ischemic heart disease	Lower-respiratory infections	Diabetes	Neonatal disorders
Brunei	4.58	4.79	4.06	4.83	2.64	5.57	3.49
Cambodia	14.03	7.29	6.71	6.32	7.32	4.35	2.94
Indonesia	16.89	13.43	11.47	12.43	9.28	10.48	5.5
Lao PDR	14.65	7.28	6.26	6.04	6.74	3.96	2.53
Malaysia	16.65	13.14	11.78	12.55	8.29	12.52	6.54
Myanmar	20.13	11.96	10.50	9.98	10.96	7.32	5.02
Philippines	11.68	12.25	10.68	11.34	8.4	9.04	3.53
Singapore	20.1	12.74	13.36	12.51	9.54	15.91	0.00
Thailand	25.98	18.7	16.78	15.27	13.74	15.79	7.62
Timor Leste	9.89	6.17	5.42	5.09	5.67	3.52	2.39
Vietnam	17.32	13.02	11.38	10.66	9.8	9.73	5.46
Global	32.31	17.45	15.06	14.58	13.09	12.69	7.19

Risk (%)
>15
15-10
10-5
<5

[a] Chronic obstructive pulmonary disease

Fig. 3.2 Percentage risk of death from specific causes attributable to ambient air pollution in each Southeast Asian nation in 2019, based on PM$_{2.5}$ and O$_3$ levels. Data from the Global Burden of Diseases—Compare Visualization (IHME 2020)

The risk of death due to ambient air pollution in Southeast Asia is generally lower than the global average. For example, the global average risk of death from lung cancer due to poor air quality is greater than 15%, but it ranges from 4% to 13% in Southeast Asian countries, with the exception of Thailand, where it reaches nearly 17%. These variations reflect not only exposures to airborne pollutants, but also other social, economic, and demographic factors that affect the underlying health and vulnerability of populations in each country to air pollution. Figure 3.2 shows the percentage risk of death from specific causes attributable to outdoor air pollution for each country in the region.

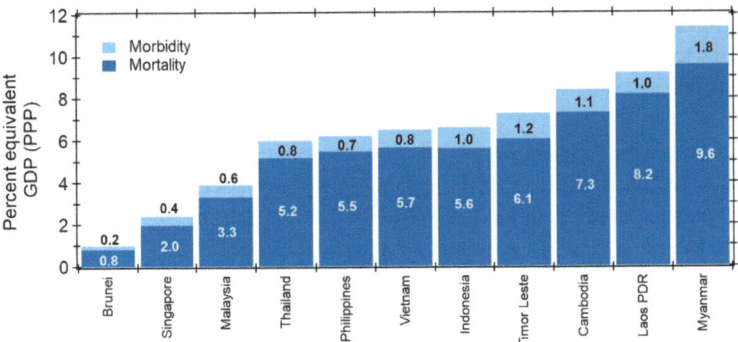

Fig. 3.3 Monetary cost of health damage from exposure to PM$_{2.5}$ pollution in 2019 for each Southeast Asian nation as determined by the World Bank (2022). The costs consider only the economic losses associated with premature mortality and morbidity, using data from the Global Burden of Diseases (HEI 2020). Costs are expressed as a percentage of GDP stated in purchasing power parity (PPP) in US$

The impact of air pollution extends beyond public health. People who become ill as a result of breathing in polluted air are more likely to miss work and have lower productivity, undermining their contributions to economic growth. Air pollution imposes a heavy economic burden on the economies of individual countries and on the region as a result of illness, premature death, lost earnings, and increased healthcare expenditure, all of which constrain productivity and economic growth. Figure 3.3 shows the monetary cost of mortality and morbidity caused by exposure to fine particles (PM$_{2.5}$) pollution for each country in the region, as determined by the World Bank (2022).

In 2019, the economic loss caused by PM$_{2.5}$ air pollution in Southeast Asia was US$506-billion in nominal terms, equivalent to 6.75% adjusted to the regional GDP stated in purchasing power parity (PPP). This cost is slightly higher than the global cost (6.1% GDP adjusted to PPP). The monetary costs ranged from 1.0% of PPP GDP in Brunei Darussalam to 11.4% in Myanmar. Unsurprisingly, the economic impact of air pollution is worse in nations with the lowest incomes. Myanmar, Lao PDR, and Cambodia were the most affected countries in the region. Premature death accounts for approximately 86% of the region's estimated economic cost, with morbidity accounting for 14%.

3.1 Air Quality Trends

The growing trend of adverse effects caused by air pollution responds to a worsening of air quality, particularly due to PM$_{2.5}$, in much of the region over the last three decades as a result of uncontrolled urbanization, rapid industrialization, increased energy consumption and motorization, and aggressive deforestation and expansive agriculture, all accompanied by a lack of strict environmental legislation. Figure 3.4 shows the trend of population-weighted ambient concentrations of ozone (O_3), $PM_{2.5}$, and nitrogen dioxide (NO_2) experienced in the region over the same period of time, as estimated in the State of Global Air 2024 Report (HEI 2024). It should be noted that these pollutant trends are based on estimates rather than field observations, as explained below, and thus may be subject to some degree of uncertainty. Similarly, a direct comparison of population-weighted concentrations to air quality guidelines or standards designed for regulatory and warning purposes is not strictly feasible due to differences in computing methods; however, such a comparison can provide insight into how close the levels of these three pollutants are to those of public health concern.

Population exposure levels to ambient concentrations are based on a combination of ground-level monitoring and chemical-transport models that estimate concentrations at a fine granular scale using data on emissions, chemical reactions, and meteorological processes, as detailed in Chapter 4.

The method assesses human exposure to O_3 in terms of the population-weighted average seasonal 8-h daily maximum concentration, taking into account the number of people who live in the area as well as the concentration to which they are exposed. Population exposure to ambient $PM_{2.5}$ and NO_2 is estimated in a similar manner, but considering population-weighted annual average concentrations and integrating satellite observations. Population-weighted average concentrations are better estimates of population exposures, because they give proportionally more weight to the air pollution experienced in areas where the majority of the population resides.

Ozone is a secondary pollutant formed through a set of chemical reactions between nitrogen oxides (NO_X) and volatile organic compounds (VOCs) in the presence of sunlight. With the exception of Brunei Darussalam and the Philippines, every nation in the region has shown an upward trend in O_3 concentrations so far this century. Brunei Darussalam and the Philippines are the only two countries that meet the air quality guideline (AQG) recommended by the World Health Organization (WHO) of 60 μg m^{-3} (~ 30 ppb at 25 °C and 1 atm) for peak O_3 ambient concentration (average of daily maximum 8-h mean O_3 concentration in the six consecutive months with the highest six-month running-average O_3 concentration). Timor-Leste barely meets interim target #2 of 70 μg m^{-3} (~ 35 ppb at 25 °C and 1 atm), while the other eight countries exceed it but fall short of interim target #1 of 100 μg m^{-3} (~ 50 ppb at 25 °C and 1 atm).

$PM_{2.5}$ typically originates from direct emissions of particles into the atmosphere as a result of combustion processes and industrial activities (primary particles), as well as indirect emissions in which particles are formed in the air from precursor pollutant species emitted by said emission sources (secondary particles). On a regional scale, direct $PM_{2.5}$ contributions from natural emission sources, such as wildfires, soil erosion, and marine aerosol, and those resulting from chemical reactions between precursor gases emitted by plants and soil (biogenic emissions) are also important.

No nation in Southeast Asia meets the WHO-recommended annual AQG level of 5 μg m^{-3} for $PM_{2.5}$. Myanmar and Thailand present the highest average annual population-weighted concentrations in the region. Both nations have achieved some

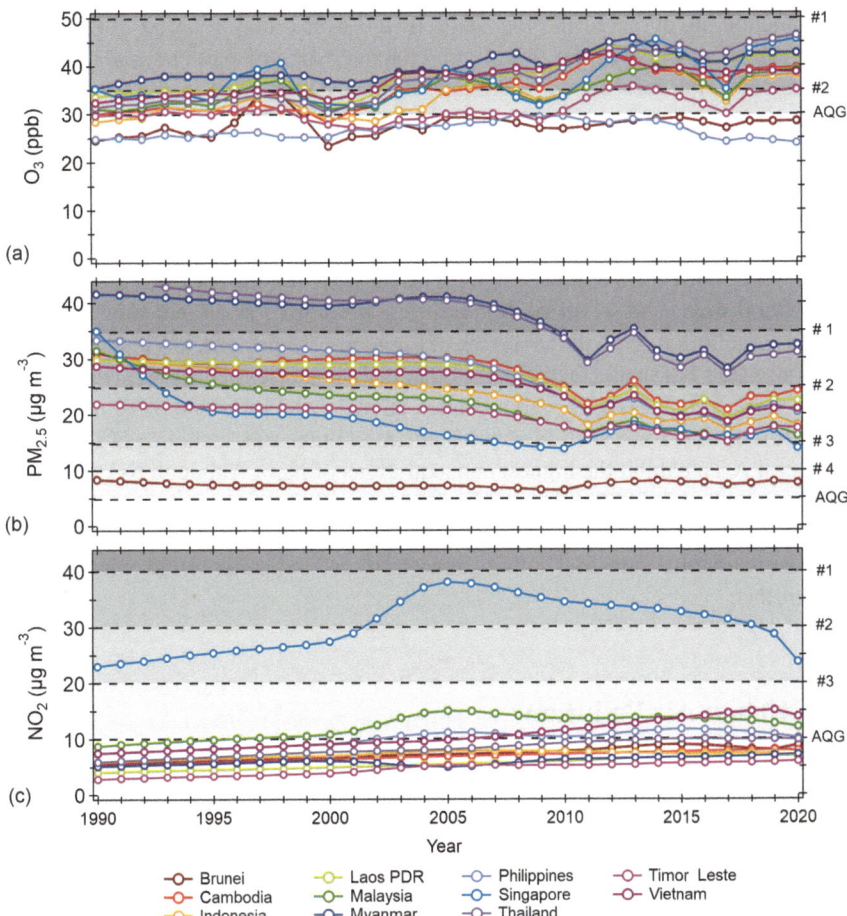

Fig. 3.4 Trends in average seasonal population-weighted ambient O_3 concentrations (**a**) and average annual population-weighted ambient concentrations of $PM_{2.5}$ and NO_2 (**b, c**) for each Southeast Asian nation. Layers of various shades of grey are used to indicate the World Health Organization's (WHO) interim targets and Air Quality Guideline (AQG) levels for both pollutant species. State of Global Air 2024 data were used to construct these charts (HEI 2024)

reduction, but continue to exceed the interim target #2 of 25 $\mu g\ m^{-3}$. Except for Brunei Darussalam and Singapore, the remaining countries exceed the interim target #3 of 15 g m^{-3}. Singapore managed to comply only by 2020, while Brunei Darussalam has never exceeded it, becoming the cleanest country in terms of particle pollution, although still above the recommended annual AQG. It is promising that the countries with the highest $PM_{2.5}$ concentrations at the turn of the century started to reduce particle pollution in the last 5–10 years. The most industrialized nations, such as Malaysia, Singapore, and possibly Indonesia, have stalled in the effort despite putting in place stricter control measures. The same is true for the case of O_3. This phenomenon, as observed in other parts of the world, may respond to increasing urbanization, changes in the origin and characteristics of particles and precursor species due to the presence of emerging emission sources, and changes in emission profiles as a result of the implementation of advanced technology and new regulatory measures; not to mention changes in regional and local meteorology triggered by a changing climate that could favor the emission, formation, and accumulation of atmospheric pollutants (Molina 2021; McDonald et al. 2018).

Nitrogen dioxide is a pollutant generated mainly by vehicles, industries, and power plants, alongside others emission sources associated with the burning of fossil fuels, agricultural waste and wildfires. In cities, it is often used as an indicator of traffic-related air pollution. With the exception of Singapore, Malaysia and Vietnam, despite showing a slow but steady increase, the other eight countries of the region meet or are very near to meet the WHO-recommended annual AQG threshold of 10 μg m^{-3}. In contrast to PM$_{2.5}$, some high-income countries in the world, including Singapore in Southeast Asia, exhibit the highest concentrations of NO$_2$. Although NO$_2$ exposure has decreased since 2005, Singapore's levels remain significantly higher than those of other nations in the region. As a city-state, it is prone to elevated concentrations, however its intense industrial activity, busy port, and vehicular traffic enhance them even further.

3.2 Urban Air Pollution

Ambient air pollution is often more severe in cities. Unhealthy air tends to be more prevalent in cities (HEI 2022). Pollutant concentrations are typically higher in cities and downwind locations than in rural or forested areas far from large urban centers. Figure 3.5 shows population-weighted concentrations of PM$_{2.5}$ and NO$_2$ for a number of selected cities of Southeast Asia, including the 25 most populated.

In terms of PM$_{2.5}$, only a few cities in the Philippines meet the interim target #4 (10 μg m^{-3}) and perhaps a dozen throughout the region meet the AQG level (5 μg m^{-3}) recommended by WHO. Except for cities in north Vietnam, Peninsular Malaysia, and Indonesia, most cities in the region meet or are close to meeting interim target #1 (35 μg m^{-3}). Cities in the lower Mekong Basin (Thailand, Loa PDR, Cambodia, and Vietnam) fall between interim targets #2 and #3 (25 and 15 μg m^{-3}, respectively), with only a few cities, like Chiang Mai and Luang Prabang, exceeding interim target #2. Indonesia's cities, particularly those in west Java and south Sumatra, which exceed interim target #4, are the most particle-polluted cities in the entire region. Population-weighted concentrations reported for Jakarta, Bandong, Malanbong, Jambi, and Palembang are above 50 μg m^{-3}, but still fall short of the extremely high concentrations reported for cities in China, India, and Sub-Saharan Africa. Cities in Malaysia, including Kuala Lumpur and Johor Bahru, as well as Singapore, have concentrations above interim target #1 as a result of their extensive urbanization and intense industrial activity, as well as periodic smoke-haze plumes originated from wildfires on neighboring Indonesian islands.

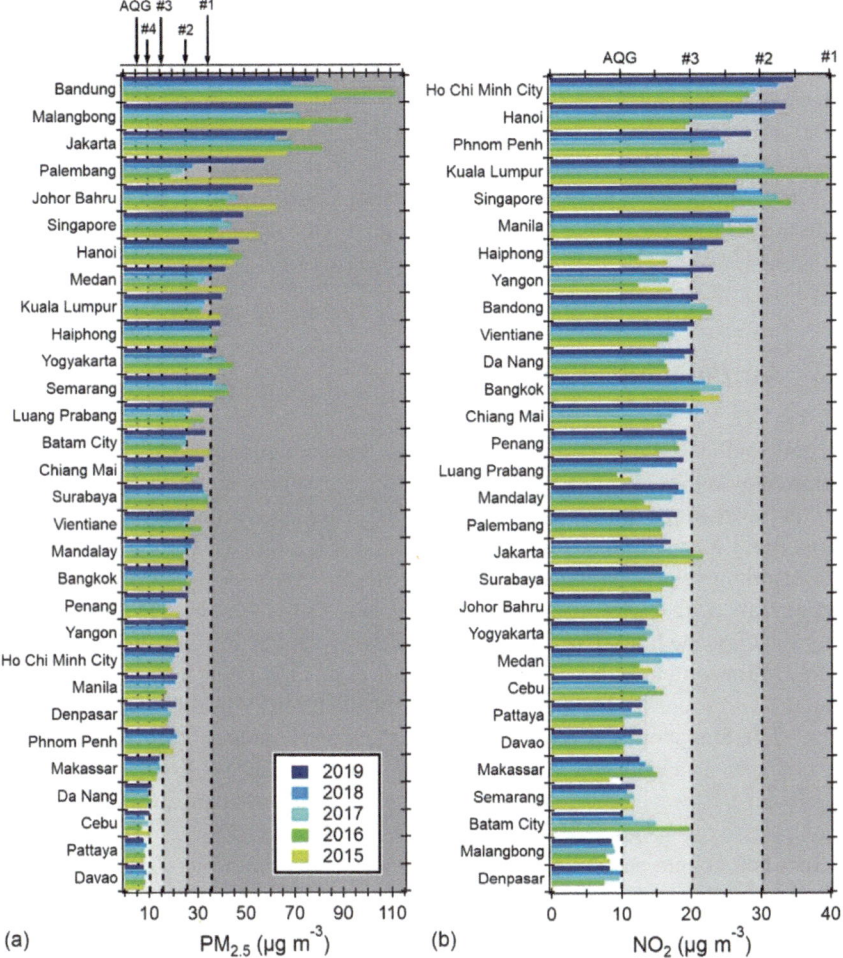

Fig. 3.5 Population-weighted annual average concentrations of $PM_{2.5}$ (**a**) and NO_2 (**b**) for 30 selected cities of Southeast Asia, including the 25 most populated. The AQG levels and interim targets for both pollutant species recommended by WHO are indicated by layers of varying shades of gray. Data obtained from the State of Global Air Report—Air Quality and Health in Cities (HEI 2022)

Regarding NO_2, it is not surprising that the highest concentrations are found in large cities with traffic problems, such as Ho Chi Minh City, Hanoi, Kuala Lumpur, and Manila, among many others. These cities all have a high proportion of motorbikes in their vehicle fleets. Motorbikes are swift, affordable, and appropriate for the urban landscape and road infrastructure of these cities. They solve mobility problems, but their noxious exhaust plumes pose a health risk. Most motorbikes in these countries lack effective emission controls, and as a result of their widespread use, they are the main source of NO_2, as well as a variety of other harmful pollutants (Velasco et al. 2022). All cities meet WHO interim target #1 (40 μg m^{-3}) but exceed interim target #2 (25 μg m^{-3}). Many cities present population-weighted concentrations between interim targets #2 and #3 (25–15 μg m^{-3}), and only a few cities meet the WHO's AQG level (10 μg m^{-3}). It is worth noting that, unlike $PM_{2.5}$, there is no interim target #4 for NO_2.

Similar to population-weighted concentrations at the national scale, concentrations at the city scale should be used as a rough reference for determining how close $PM_{2.5}$ and NO_2 levels are to public health concern levels.

3.3 Air Pollutants Emission

In rural Southeast Asia, biomass burning is the primary emission source of pollutants, followed by vehicle exhaust, whereas in urban areas, vehicle and industrial emissions are the most important sources of air pollution. Open burning of municipal and agricultural waste is banned in most countries, yet it is still a common practice.

Regarding vehicle emissions, most countries have set stricter emission limits in recent years, with emission standards equivalent to Euro IV for both heavy and light-duty vehicles, and Euro III or IV for motorcycles, in conjunction with the introduction of low sulfur content fuels (50 ppm, see Table 3.1). Countries that have not yet done so plan to do so soon. Thailand has already established Euro V standards for passenger cars, while Singapore has gone a step further and has established Euro VI standards for both diesel and gasoline vehicles, as well as has introduced ultra-low sulfur fuels ($\leq$ 10 ppm). However, three countries still lack emission standards, while Cambodia only has one for passenger cars.

In some countries, the lack of equipment and legislation to regularly test the exhaust emissions of older cars is a major problem. The absence of regulations on second-hand vehicle imports aggravates the problem since it opens the door to less fuel-efficient and more polluting vehicles.

Eight countries in the region present low industrial energy efficiency as measured by GDP per unit of energy; Singapore, Malaysia, and Thailand are more energy efficient. This suggests that the industrial technology used in the former countries is outdated, resulting in energy waste and unnecessary emission of pollutants. In general, low-efficiency industries emit more pollutants, both directly and indirectly, than industries equipped with more efficient technologies (UNEP 2017). To make this situation worse, Indonesia and Lao PDR plan to increase the share of coal in electricity generation, and Timor-Leste has purchased and relocated second-hand heavy oil power plants from China.

The worst air pollution episodes in the region are caused by massive wildfires associated with aggressive deforestation and expansive agriculture. The oil palm and pulpwood agroindustry in the Indonesian's islands of Kalimantan and Sumatra are the culprits of almost annual smoke-haze (local term for large-scale plumes of pollutants from wildfires) events (Field et al. 2009). These air pollution events are

Table 3.1 Standards for vehicle emissions and sulfur levels in fuels. Data obtained from government websites, Clean Air Asia (2019), and ASEAN (2019)

Country	Heavy duty vehicles	Light-duty vehicles	Motorcycles	Sulfur content in diesel (ppm)	Sulfur content in gasoline (ppm)
Brunei Darussalam	–	–	–	50	50
Cambodia	Euro IV	Euro IV	Euro III	50	50
Indonesia	Euro IV	Euro IV	Euro IV	2000	500
Lao PDR	–	Euro II	–	2500	–
Malaysia	Euro IV	Euro IV	Euro IV	50	50
Myanmar	–	–	–	–	–
Singapore	Euro VI	Euro VI	Euro VI	10	10
Philippines	Euro IV	Euro IV	Euro III	50	50
Thailand	Euro III	Euro V	Euro IV	50	50
Timor-Leste	–	–	–	500	150
Vietnam	Euro IV	Euro IV	Euro III	500	500

capable of blanketing the sky of Indonesia, Malaysia, and Singapore for days, if not weeks (Reddington et al. 2014). In 2015, the Indian Ocean Dipole (IOD) and *El Niño* Southern Oscillation (ENSO) came together and produced drought conditions in the Indonesian archipelago, triggering a smoke-haze episode that lasted more than a month and is estimated to have resulted in approximately 100-thousand premature deaths (Koplitz et al. 2016). In the lower Mekong Basin, smoke-haze episodes caused by agricultural waste burning during the dry season can sometimes have a severe impact on air quality (Khodmanee and Amnuaylojaroen 2021; Huang et al. 2013).

Figure 3.6 shows each country's emissions contribution for a set of selected pollutant species as estimated by EDGARv8.1—Emissions Database for Global Atmospheric Research using 2022 as reference year (https://edgar.jrc.ec.europa.eu/). This figure also shows the total greenhouse gas (GHG) emissions (CO_2, CH_4, N_2O, and fluorinated gases) as estimated by EDGARv8.0 (Crippa et al. 2024). Figure 3.7 shows the emissions evolution for seven of the pollutants shown in Fig. 3.6 from 1970 to 2015. Please keep in mind that the emission data presented here is subject to uncertainties. Global emission datasets, as explained further in Sect. 6.1, fill the gap in emission inventories constructed at the regional or country scale but are subject to errors due to the use of generic emission factors and a lack of detailed data for various economic activities.

Indonesia, as the region's largest economy and most populous nation, has the highest emission contribution for all pollutant species evaluated by EDGAR. Depending on the pollutant, Indonesia's emissions account for 33-50% of Southeast Asia's total emissions budget. Vietnam's contributions range from 11% to 20%, depending on the pollutant, becoming the second largest contributor. Thailand follows closely behind, contributing between 8% and 17%. Despite having a

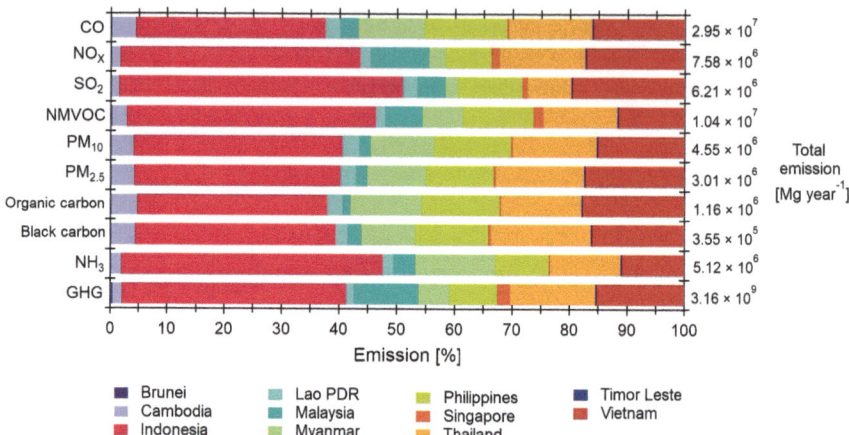

Fig. 3.6 Percentage of emission contributions for selected air pollutants for each country in Southeast Asia as estimated in EDGARv8.1—Emissions Database for Global Atmospheric Research for 2022 (https://edgar.jrc.ec.europa.eu/), and for total GHG (CO_2, CH_4, N_2O, and fluorinated gases) expressed as CO_2 equivalent as estimated by EDGARv8.0 (Crippa et al. 2024). On the right side of the graph, the total emissions of each pollutant species for the entire region are displayed

population 30% smaller, Thailand's economy is 30% stronger than Vietnam's. The Philippines' emissions are next, with contributions ranging from 8% to 14%. Despite having the weakest economy in terms of GDP per capita, Myanmar ranks fifth. Pollutant emissions from agriculture and biomass burning, such as ammonia (NH_3), CO, black carbon, and organic aerosols, are particularly high in Myanmar.

Malaysia and Singapore have achieved a high level of industrialization, as evidenced by their relatively high emissions of pollutant species associated with industrial activity and the use of fossil fuels, such as NO_X, non-methane VOCs (NMVOC), and GHG. However, Malaysia's emissions are significantly higher than Singapore's by a factor of five for GHG and black carbon and up to eleven for CO, which is inconsistent with the income difference between both countries. Singapore's nominal GDP is only 7% higher than Malaysia's, but the difference in per capita terms is six times larger. This is in response to different types of industries in both countries and perhaps to Singapore's greater progress in controlling industrial emissions. Brunei Darussalam's net emissions are low, but in per capita terms are among the highest, especially of pollutants associated with industrial activity such as NO_X, NMVOC, and CO. Emissions from Lao PDR and Cambodia are small, while those from Timor-Leste are even smaller. The combined emissions of these three countries represent between 3% and 7% of the total emissions depending on the pollutant and are in line with their low economic development.

Southeast Asia's rapid economic growth over the last 50 years has consequently resulted in a significant increase in the emission of atmospheric pollutants. As shown in Fig. 3.7, emissions increased by 150–500% between 1970 and 2015, depending on the pollutant species. The momentum of emissions growth for some pollutant

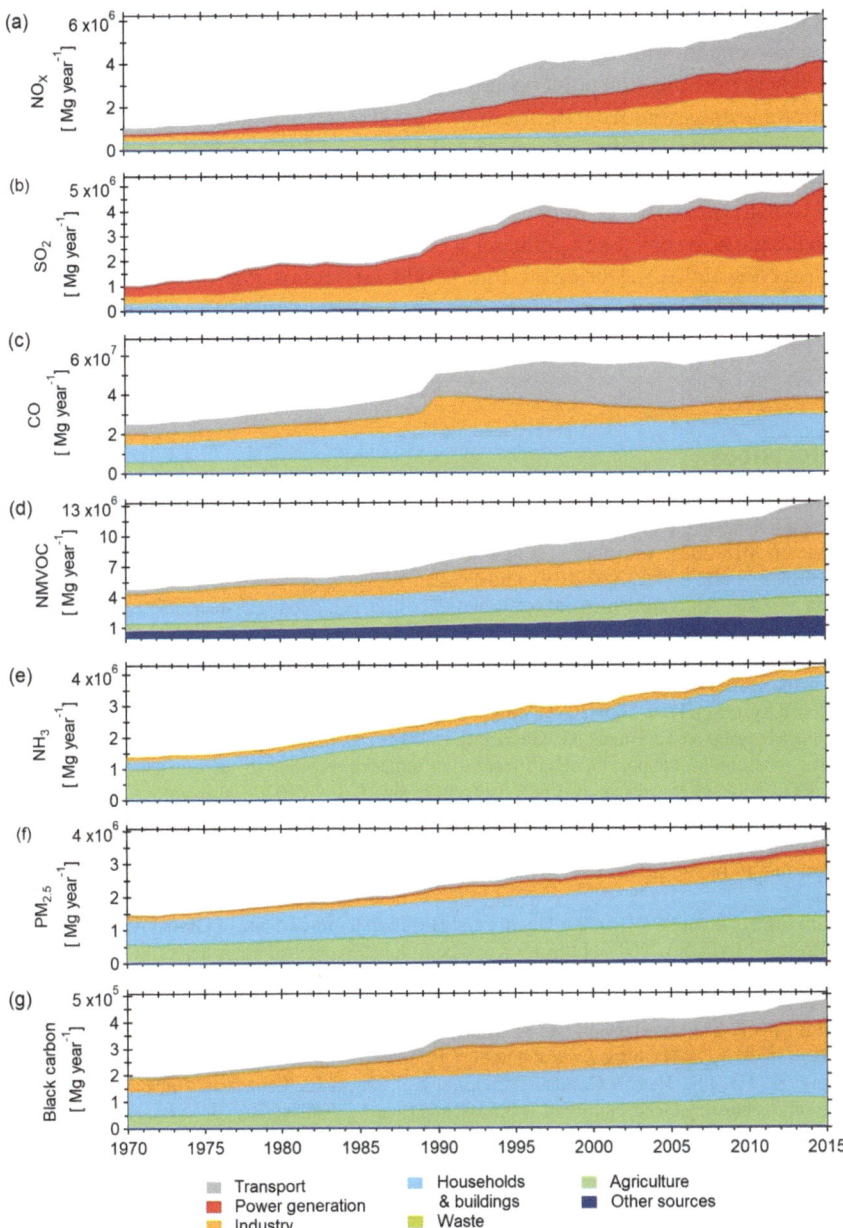

Fig. 3.7 Emission trends of NO_X ($NO_X = NO + NO_2$), SO_2, CO, non-methane volatile organic compounds (NMVOC), ammonia (NH_3), $PM_{2.5}$, and black carbon in Southeast Asia. Data obtained from EDGARv5.0—Emissions Database for Global Atmospheric Research (https://edgar.jrc.ec.eur opa.eu/, Crippa et al. 2020)

species associated with industrial activity began to slow in the 1990s, but not for those associated with transportation. This phenomenon is evident in the case of CO, where the inflection point occurred in 1990, when industrial emissions began to decline and emissions from the transportation sector began to contribute significantly more. In the case of pollutants associated with agriculture, such as ammonia (NH_3), there has been a steady increase in emissions since then.

From the turn of the century to 2015, emissions increased between 20% and 40%, with the greatest increase in pollutant species associated with fossil fuel combustion, such as NO_X and SO_2. During this time period, GHG emissions increased by 39% due to an increase in fossil fuel consumption. From 2010 to 2015, all pollutant emissions increased by 10–15%.

References

Association of Southeast Asian Nations (ASEAN): ASEAN fuel economy roadmap for the transport sector 2018-2025: with focus on light-duty vehicles. ASEAN Secretariat Jakarta, Indonesia. ISBN 978-602-5798-32-0 (2019). https://asean.org

Clean air Asia: Air quality in Asia: status and trends (2019). https://cleanairasia.org/our-resources

Crippa, M., Guizzardi, D., Pagani, F., Schiavina, M., Melchiorri, M., Pisoni, E., Graziosi, F., Muntean, M., Maes, J., Dijkstra, L., Van Damme, M., Clarisse, L., and Coheur, P.: Insights on the spatial distribution of global, national and sub-national GHG emissions in EDGARv8.0, Earth Syst. Sci. Data 16(6), 2811-2830 (2024). https://doi.org/10.5194/essd-16-2811-2024

Crippa, M., Solazzo, E., Huang, G., Guizzardi, D., Koffi, E., Muntean, M., Schieberle, C., Friedrich, R., Janssens-Maenhout, G.: High resolution temporal profiles in the emissions database for global atmospheric research. Earth Syst. Sci. Data 7, 121 (2020). https://doi.org/10.1038/s41 597-020-0462-2

Field, R.D., van der Werf, G.R., Shen, S.S.P.: Human amplification of drought-induced biomass burning in Indonesia since 1960. Nat. Geosci. 2, 185–188 (2009). https://doi.org/10.1038/nge o443

Health Effects Institute (HEI): Air Quality and Health in Cities: a State of Global Air Report 2022. Boston, MA, USA (2022). https://www.stateofglobalair.org/resources/health-in-cities

Health Effects Institute (HEI): State of Global Air 2020. Data source: Global Burden of Disease Study 2019. IHME (2020). https://www.stateofglobalair.org/

Health Effects Institute (HEI): State of Global Air 2024. Special Report. Health Effects Institute, Boston, MA (2024). https://www.stateofglobalair.org/

Huang, K., Fu, J.S., Hsu, N.C., Gao, Y., Dong, X., Tsay, S.C., Lam, Y.F.: Impact assessment of biomass burning on air quality in Southeast and East Asia during BASE-ASIA. Atmos. Environ. 78, 291–302 (2013). https://doi.org/10.1016/j.atmosenv.2012.03.048

Institute for Health Metric and Evaluation (IHME).: Global burden of diseases 2019. University of Washington (2020). https://www.healthdata.org/gbd/2019

Institute for Health Metrics and Evaluation (IHME).: Global burden of diseases compare data visualization. University of Washington, Seattle, WA, IHME, (2024). Available from https://viz hub.healthdata.org/gbd-compare/

Khodmanee, S., Amnuaylojaroen, T.: Impact of biomass burning on ozone, carbon monoxide, and nitrogen dioxide in Northern Thailand. Front. Environ. Sci. 9, 27 (2021). https://doi.org/10. 3389/fenvs.2021.641877

Koplitz, S.N., Mickley, L.J., Marlier, M.E., Buonocore, J.J., Kim, P.S., Liu, T., Sulprizio, M.P., DeFries, R.S., Jacob, D.J., Schwartz, J., Pongsiri, M., Myers, S.S.: Public health impacts of the severe haze in Equatorial Asia in September–October 2015: demonstration of a new framework for informing fire management strategies to reduce downwind smoke exposure. Environ. Res. Lett. **11**, 094023 (2016). https://doi.org/10.1088/1748-9326/11/9/094023/meta

McDonald, B.C., De Gouw, J.A., Gilman, J.B., Jathar, S.H., Akherati, A., Cappa, C.D., Jimenez, J.L., Lee-Taylor, J., Hayes, P.L., McKeen, S.A., Cui, Y.Y., Kim, S.W., Gentner, D.R., Isaacman-Vanwertz, G., Goldstein, A., Harley, R.A., Frost, G.J., Roberts, J.M., Ryerson, T.B., Trainer, M.: Volatile chemical products emerging as largest petrochemical source of urban organic emissions. Science **359**(6377), 760–764 (2018). https://doi.org/10.1126/science.aaq0524

Molina, L.T.: Introductory lecture: air quality in megacities. Faraday Discuss. **226**, 9–52 (2021). https://doi.org/10.1039/D0FD00123F

Reddington, C.L., Yoshioka, M., Balasubramanian, R., Ridley, D., Toh, Y.Y., Arnold, S.R., Spracklen, D.V.: Contribution of vegetation and peat fires to particulate air pollution in Southeast Asia. Environ. Res. Lett. **9**, 094006 (2014). https://doi.org/10.1088/1748-9326/9/9/094006

United Nations Environment Programme.: Southeast Asia: Actions Taken by Governments to Improve Air Quality. United Nations (2017). https://wedocs.unep.org/20.500.11822/31913

Velasco, E., Ha, H.H., Pham, A.D., Rastan, S.: Effectiveness of wearing face masks against traffic particles on the streets of Ho Chi Minh City, Vietnam. Environ. Sci.: Atmos. (2022). https://doi.org/10.1039/D2EA00071G

World Bank: The Global Health Cost of $PM_{2.5}$ Air Pollution: a Case for Action Beyond 2021. In: International Development in Focus. World Bank, Washington, DC (2022). https://openknowledge.worldbank.org/handle/10986/36501

Chapter 4
Air Quality Data Sources

Abstract This chapter describes the methodology and data sources used for the air quality assessments presented in the previous chapter. The State of Global Air report is presumably the most complete source of information available about the current state of air quality in Southeast Asia. It provides a comprehensive analysis of air quality levels and trends, as well as associated health impacts, for each country. It characterizes $PM_{2.5}$ and O_3 concentrations at a spatial resolution useful for informed action by combining ambient concentrations measured by ground-level monitors, data retrieved from multiple satellite instruments, and the outputs of chemical transport models. Its results are used to estimate exposure population-weighted average concentrations, as well as the associated health damages, mortality risk, and monetary cost. However, the limited knowledge on the characteristics of local air pollution, the small number of air quality monitoring stations, reduced satellite sampling due to frequent overcast skies, and the representativeness of the chemical transport models' algorithms for the region's conditions all jeopardize the estimates' accuracy.

Keywords State of Global Air · Air quality and health in cities · Population-weighted average concentration · Air pollution health cost

The information on air pollution presented in the preceding section is the result of collaborative and multidisciplinary efforts by international consortia committed to collecting and analyzing the largest amount of available data to characterize the current state of air quality and potential impacts on a global and regional scale. In the case of Southeast Asia, the products of these consortia represent the most up-to-date and complete source of information on the subject at the time of writing. The information provided by these international assessments can generally be used as a starting point for specific assessments or as reference data for academic studies. However, due to uncertainty in the inputs, it is recommended that they be used for regulatory and policy purposes only in a qualitative manner in the absence of another source of information. The accuracy of the information they provide depends on the proper operation of the models and methodologies employed. The performance of the models depends on the availability of reliable data on the parameters necessary

© The Author(s), under exclusive license to Springer Nature Switzerland AG 2024   37
E. Velasco et al., *Air Quality Management and Research in Southeast Asia*,
SpringerBriefs in Earth System Sciences, https://doi.org/10.1007/978-3-031-69088-4_4

for their operation and the evaluation of their results (e.g., meteorological and air quality data at ground level), as well as the representativeness of their algorithms for the conditions of the region of concern. Also, keep in mind that these products have been developed to characterize air pollution under regular circumstances, not during specific events (e.g., smoke-haze episodes triggered by wildfires).

This chapter provides a concise description of the methodology and data sources utilized for such air quality assessments, as well as a discussion of the associated uncertainties. The products derived from them, the data sources used for their construction, and other pertinent sources for air quality data sources are listed in Table 4.1.

The following chapters delve into each aspect of air quality management, including air quality monitoring, emission inventories, numerical modeling, emerging methodologies, such as satellite remote sensing, and scientific research. Readers are also encouraged to read two recent articles that provide a brief overview of the current state of air quality and related policies in various countries and cities throughout the region. Oanh et al. (2023) provide an overview of air pollution in Thailand, Vietnam, and Myanmar, whereas Cambaliza et al. (2023) concentrate on air quality in Manila, Kuala Lumpur, Singapore, and Jakarta.

4.1   State of Global Air

The State of Global Air is a report and interactive website developed by the Health Effects Institute (HEI, https://www.healtheffects.org/) that provides a comprehensive analysis of the levels and trends in air quality and health for every country in the world. It is designed to provide access to reliable and meaningful information about air pollution exposure and its health effects. It is open to the public and free of charge.

The data used in the State of Global Air report is part of the Global Burden of Diseases (GBD), Injuries, and Risk Factors project, which is coordinated by the Institute for Health Metrics and Evaluation (IHME, https://www.healthdata.org/). The HEI and the School of Population and Public Health at the University of British Columbia have developed the methods and datasets necessary to estimate the disease burden resulting from outdoor air pollution.

The State of Global Air Report focuses on exposure to $PM_{2.5}$, NO_2 and O_3, the three pollutant species that pose the greatest threat to public health. $PM_{2.5}$ is the most consistent and robust predictor of mortality in studies of long-term exposure (Chen et al. 2008; Hoek et al. 2013), whereas O_3 is associated with respiratory diseases independently of $PM_{2.5}$ exposure (Jerret et al. 2009). Nitrogen dioxide is a key tracer of traffic-related air pollution. It can react with other precursor gases to produce O_3 and aerosols. Exposure to NO_2 can develop and aggravate asthma symptoms, impair lung development, increase susceptibility to respiratory infections, and exacerbate allergies (Health Canada, 2016; US-EPA, 2016). Furthermore, substantial evidence supports the association between long-term exposure to NO_2 and mortality (HEI, 2022).

Table 4.1 Air quality information products and data sources for Southeast Asia

Product/Dataset	Information/Data	Scale	Countries covered	Reference
State of Global Air (HEI 2024)	Annual trends in air quality and health effects. Exposure concentrations to $PM_{2.5}$, NO_2 and O_3	Country and subnational for selected countries	Every country in the world	https://www.stateofglobalair.org/
Air Quality and Health in Cities—State of Global Air (HEI 2022)	Annual trends in air quality and health effects. Exposure concentrations to $PM_{2.5}$ and NO_2	City	7,239 cities across the world, 516 cities in Southeast Asia	https://www.stateofglobalair.org/resources/health-in-cities
The Global Health Cost of $PM_{2.5}$ Air Pollution (World Bank 2022)	Monetary valuation of the global cost of mortality and morbidity caused by exposure to ambient $PM_{2.5}$	Country	Every country in the world	https://openknowledge.worldbank.org/handle/10986/36501
WHO Air Quality Database (WHO 2022)	Annual ambient concentrations of NO_2, PM_{10} and $PM_{2.5}$ at urban scale	City	Cities from 117 countries worldwide	https://www.who.int/data/gho/data/themes/air-pollution
Tropospheric Ozone Assessment Report (Schultz et al. 2017)	A suite of O_3 data products including standard statistics, health and vegetation impact metrics, and trend information	Individual stations	Worldwide	https://toar-data.org
AirNow-DOS Program	Hourly ambient concentrations of $PM_{2.5}$	Individual stations	US embassies and consulates around the world	https://www.airnow.gov/international/us-embassies-and-consulates/

(continued)

Table 4.1 (continued)

Product/Dataset	Information/Data	Scale	Countries covered	Reference
Acid Deposition Monitoring Network in East Asia (EANET) (Akimoto et al. 2022)	Wet and dry deposition of anions (SO_4^{2-}, NO_3^-, Cl^-, F^-, PO_4^{3-}), cations (Na^+, K^+, NH_4^+, Ca_2^+, Mg^{2+}, HCO_3^+), and organic acids, as well as measurements of gaseous species (SO_2, HNO_3, HCl, NH_3) using passive samplers	Individual stations	66 sites (26 urban, 19 rural, 21 remote) throughout East Asia	https://www.eanet.asia/
Surface Particulate Matter Network (SPARTAN) (Snider et al. 2015, 2016)	$PM_{2.5}$ mass concentration, chemical composition, and optical characteristics	Individual stations	29 sites around the world	https://www.spartan-network.org/
AEronet RObotic NETwork (AERONET) (Holben et al. 1998)	Aerosol optical depth	Individual stations	> 600 sites worldwide	https://aeronet.gsfc.nasa.gov/
Pandonia Global Network (Szykman et al. 2019)	Total column and vertically resolved concentrations of a suite of pollutant species, focusing on O_3, NO_2, and HCHO	Individual stations	> 120 sites worldwide	https://www.pandonia-global-network.org/
OpenAQ	Ambient concentrations of PM_{10}, $PM_{2.5}$, O_3, SO_2, CO, NO_2, and black carbon	Individual stations	Worldwide	https://openaq.org/

Ambient concentrations of $PM_{2.5}$ are calculated using a Bayesian hierarchical framework comprised of ambient $PM_{2.5}$ measurements from over 10,000 air quality monitors, aerosol optical depth (AOD) data from multiple satellite instruments (the Moderate Resolution Imaging Spectroradiometer, SeaWiFs, and the Multiangle Imaging Spectroradiometer), and $PM_{2.5}$ predictions from the GEO-Chem chemical transport model across the globe in grid cells of 0.1° × 0.1° of longitude and latitude (approximately 11 km × 11 km at the equator), for details see Shaddick et al.

(2018). This combined approach yields annual averages, which are then linked to the number of people living in the corresponding grid cells to produce a population-weighted annual average concentration, and thus the concentration that a population is most likely to encounter in a given location. The gridded exposure concentrations are aggregated to national-level population-weighted means with a 95% confidence interval, which is derived by sampling 1000 draws of each grid cell value and its uncertainty distribution.

Similar to what is done for $PM_{2.5}$, data from more than 8,800 ground-level air quality monitors and outputs from nine atmospheric chemical transport models are combined to characterize O_3 concentrations and trends across the globe in grid cells of 1 km × 1 km (Brauer et al. 2016). Similarly, the ambient concentrations of O_3 are linked to the number of inhabitants in each analyzed cell, but taking into account seasonal (summer) 8-h daily maximum concentrations. Consequently, the method calculates population-weighted average seasonal 8-h daily maximum O_3 concentration. Regarding NO_2, annual average exposure estimates are derived using ground-based measurements from 5,220 monitors in 58 countries, land-use regression model predictions, and satellite-based input data.

The risk of mortality associated with O_3 and $PM_{2.5}$ pollution is then estimated for each cause of death, including ischemic heart disease, cerebrovascular disease, COPD, lung cancer, and lower respiratory infections, using integrated exposure–response functions from epidemiology studies. Cohen et al. (2017) provide a detailed description of the methodology used to estimate exposure population-weighted average concentrations and associated mortality risks. For the case of NO_2, the report looks into the development of childhood asthma.

The satellite and modeling approaches have proven to be reasonably accurate when compared to ground-level measurements. The resulting $PM_{2.5}$ ambient concentrations from the combined approach used in the State of Global Air report show a strong correlation ($r^2 = 0.81$) with concentrations from air quality monitors around the world (van Donkelaar et al. 2016). However, in regions like Southeast Asia, where air quality monitors are scarce, as are ground-based solar photometers needed to calibrate the algorithms used to retrieve AOD from satellites, and knowledge of the optical properties of the aerosols is limited, the approach's performance is affected. Another issue to consider is reduced satellite sampling due to the frequent presence of cirrus clouds covering the sky.

Another factor to consider is the lack of information on economic activities and demographic parameters in some regions of Southeast Asia. It makes it difficult to accurately estimate pollutant emissions, jeopardizing the outcomes of models used to evaluate the transport and transformations of pollutants from their emission sources to receptor sites.

Taking into account all sources of uncertainty, van Donkelaar et al. (2016) reported a mean uncertainty of 6 μg m^{-3} (bias at regional scale) in estimated concentrations of $PM_{2.5}$ (26.0 μg m^{-3}) compared to concentrations measured in situ at 62 locations in Southeast Asia (27.2 μg m^{-3}). This bias ranked among the highest in the world. In high-income regions with dense air quality monitoring networks and ground-based solar photometers, robust information on the optical aerosol properties, and reliable emission inventories, such as Europe and North America, the regional bias never exceeded $1.0\ \mu g\ m^{-3}$.

4.2   Air Quality and Health in Cities—State of Global Air

This report is an extension of the State of Global Air report (https://www.stateofgloba lair.org/resources/health-in-cities). It focuses on air pollution and its health impacts on cities all over the world, including 516 cities in Southeast Asia. It assesses exposure concentrations to $PM_{2.5}$ and NO_2, but not to O_3, using data from the GBD project.

Ambient and exposure concentrations were calculated in the same manner as described above for the State of Global Air report, but with a higher spatial resolution to reach grid cells of $0.0083° \times 0.0083°$ (approximately 1 km × 1 km at the equator) to better match the resolution of urban spatial extents (Hammer et al. 2020). Anenberg et al. (2022) and Southerland et al. (2022) provide details on $PM_{2.5}$ and NO_2 exposure concentration estimates, respectively.

The accuracy of ambient $PM_{2.5}$ and NO_2 concentrations in cities was threatened by the same factors that threatened the accuracy of ambient $PM_{2.5}$ and O_3 concentrations for the entire globe. However, advances in satellite instrumentation and AOD retrieval algorithms, improvements in GEOs-Chem simulations, including new schemes for dust and secondary organic aerosol chemistry, updated emission inventories, a larger number of ground-level monitors, and improvements in the combined methodology improved the correlation ($r^2 = 0.91$) between estimated annual average concentrations and measured concentrations *in situ* (Hammer et al. 2020).

4.3   The Global Health Cost of $PM_{2.5}$ Air Pollution

This World Bank report presents estimates of the global, regional, and national costs of health damage caused by $PM_{2.5}$ air pollution using 2019 as the reference year (World Bank 2022). It uses data from the GBD project. Estimates of the number of premature deaths and diseases in a country as a result of the population's exposure to given concentrations of $PM_{2.5}$ are used to calculate the costs associated with premature deaths and morbidity. The value of statistical life, a measure of how much individuals are willing to pay for a reduction in the risk or likelihood of premature death, is used to estimate the cost of premature death. The cost of morbidity is estimated based on the GBD project's estimation of years lived with disability. It reflects the duration and severity of diseases as a measure of disease burden. The years lived with disability due to $PM_{2.5}$ exposure are converted to days lived with disease, with the cost of a day of disease equaling a fraction of the average daily wage rate.

As in previous cases, the accuracy of the exposure concentrations calculated in the State of the Global Air, as well as the reliability of the exposure-functions used in the

GBD project to calculate the risk of health damage, determines the representativeness of the estimated health cost.

References

Akimoto, H., Sato, K., Sase, H., Dong, Y., Hu, M., Duan, L., Sunwoo, Y., Suzuki, K., Tang, X.: Development of science and policy related to acid deposition in East Asia over 30 years. Ambio **51**(8), 1800–1818 (2022). https://doi.org/10.1007/s13280-022-01702-6

Anenberg, S.C., Mohegh, A., Goldberg, D.L., Kerr, G.H., Brauer, M., Burkart, K., Hystad, P., Larkin, A., Wozniak, S., Lamsal, L.: Long-term trends in urban NO_2 concentrations and associated pediatric asthma incidence: estimates from global datasets. Lancet Planet. Health **6**(1), 49–58 (2022). https://doi.org/10.1016/S2542-5196(21)00255-2

Brauer, M., Freedman, G., Frostad, J., van Donkelaar, A., Martin, R.V., Dentener, F., van Dingenen, R., Estep, K., Amini, H., Apte, J.S., Balakrishnan, K., et al.: Ambient air pollution exposure estimation for the global burden of disease 2013. Environ. Sci. Technol. **50**(1), 79–88 (2016). https://doi.org/10.1021/acs.est.5b03709

Cambaliza, M.O.L., NUS AQ Lab., Latif, M.T., Lestari, P.: Regional and urban air quality in Southeast Asia: Maritime Continent. In: Akimoto, H., Tanimoto, H. (eds.) Handbook of Air Quality and Climate Change. Springer, Singapore (2023). https://doi.org/10.1007/978-981-15-2527-8_68-1

Chen, H., Goldberg, M., Villeneuve, P.J.: A systematic review of the relation between long-term exposure to ambient air pollution and chronic diseases. Rev. Environ. Health **23**(4), 243–298 (2008). https://doi.org/10.1515/REVEH.2008.23.4.243

Cohen, A.J., Brauer, M., Burnett, R., Anderson, H.R., Frostad, J., Estep, K., Balakrishnan, K., Brunekreef, B., Dandona, L., Dandona, R., Feigin, V., et al.: Estimates and 25-year trends of the global burden of disease attributable to ambient air pollution: an analysis of data from the Global Burden of Diseases Study 2015. Lancet **389**(10082), 1907–1918 (2017). https://doi.org/10.1016/S0140-6736(17)30505-6

Hammer, M.S., van Donkelaar, A., Li, C., Lyapustin, A., Sayer, A.M., Hsu, N.C., Levy, R.C., Garay, M.J., Kalashnikova, O.V., Kahn, R.A., Brauer, M., et al.: Global estimates and long-term trends of fine particulate matter concentrations (1998–2018). Environ. Sci. Technol. **54**(13), 7879–7890 (2020). https://doi.org/10.1021/acs.est.0c01764

Health Canada.: Human health risk assessment for ambient nitrogen dioxide. Health Canada, Ottawa (2016). https://www.canada.ca/en/health-canada/services/publications/healthy-living/human-health-risk-assessment-ambient-nitrogen-dioxide.html

Health Effects Institute (HEI).: Air quality and health in cities: A state of global air report 2022. Boston, MA, USA (2022). https://www.stateofglobalair.org/resources/health-in-citie

Health Effects Institute (HEI).: State of Global Air 2024. Special Report. Health Effects Institute, Boston, MA (2024). https://www.stateofglobalair.org/

Health Effects Institute (HEI).: Systematic review and meta-analysis of selected health effects of long-term exposure to traffic-related air pollution. Special Report 23, Health Effects Institute. Boston, MA (2022). https://www.healtheffects.org/publication/systematic-review-and-meta-analysis-selected-health-effects-long-term-exposure-traffic

Hoek, G., Krishnan, R.M., Beelen, R., Peters, A., Ostro, B., Brunekreef, B., Kaufman, J.D.: Long-term air pollution exposure and cardio-respiratory mortality: a review. Environ. Health **12**, 43 (2013). https://doi.org/10.1186/1476-069X-12-43

Holben, B.N., Eck, T.F., Slutsker, I., Tanré, D., Buis, J.P., Setzer, A., Vermote, E., Reagan, J.A., Kaufman, Y., Nakajima, T., Lavenu, F., Jankowiak, I., Smirnov, A.: AERONET—A federated instrument network and data archive for aerosol characterization. Remote Sens. Environ. **66**(1), 1–16 (1998). https://doi.org/10.1016/S0034-4257(98)00031-5

Jerrett, M., Burnett, R.T., Pope, C.A., III., Ito, K., Thurston, G., Krewski, D., Shi, Y., Calle, E., Thun, M.: Long-term ozone exposure and mortality. N. Engl. J. Med. **360**, 1085–1095 (2009). https://doi.org/10.1056/NEJMoa0803894

Oanh, N.T.K., Hlaing, O.M.T., Hien, T.T.: Regional and urban air quality in Mainland Southeast Asia countries. In: Akimoto, H., Tanimoto, H. (eds.) Handbook of Air Quality and Climate Change. Springer, Singapore (2023). https://doi.org/10.1007/978-981-15-2527-8_69-1

Schultz, M.G., Schröder, S., Lyapina, O., Cooper, O.R., Galbally, I., Petropavlovskikh, I., Von Schneidemesser, E., Tanimoto, H., Elshorbany, Y., Naja, M., Seguel, R.J.: Troposheric ozone assessment report: Database and metrics data of global surface ozone observations. Elem.: Sci. Anth. **5**, 58 (2017). https://doi.org/10.1525/elementa.244

Shaddick, G., Thomas, M.L., Amini, H., Broday, D., Cohen, A., Frostad, J., Green, A., Gumy, S., Liu, Y., Martin, R.V., Pruss-Ustun, A., Simpson, D., van Donkelaar, A., Brauert, M.: Data integration for the assessment of population exposure to ambient air pollution for global burden of disease assessment. Environ. Sci. Technol. **52**(16), 9069–9078 (2018). https://doi.org/10.1021/acs.est.8b02864

Snider, G., Weagle, C.L., Martin, R.V., Van Donkelaar, A., Conrad, K., Cunningham, D., Gordon, C., Zwicker, M., Akoshile, C., Artaxo, P., et al.: SPARTAN: a global network to evaluate and enhance satellite-based estimates of ground-level particulate matter for global health applications. Atmos. Meas. Tech. **8**, 505–521 (2015). https://doi.org/10.5194/amt-8-505-2015

Snider, G., Weagle, C.L., Murdymootoo, K.K., Ring, A., Ritchie, Y., Stone, E., Walsh, A., Akoshile, C., Anh, N.X., Balasubramanian, R., Brook, J., et al.: Variation in global chemical composition of PM$_{2.5}$: Emerging results from SPARTAN. Atmos. Chem. Phys **16**(15), 9629–9653 (2016). https://doi.org/10.5194/acp-16-9629-2016

Southerland, V.A., Brauer, M., Mohegh, A., Hammer, M.S., van Donkelaar, A., Martin, R.V., Apte, J.S. and Anenberg, S.C.: Global urban temporal trends in fine particulate matter (PM$_{2.5}$) and attributable health burdens: estimates from global datasets. Lancet Planet. Health **6**(2), 139–146 (2022). https://doi.org/10.1016/S2542-5196(21)00350-8

Szykman, J., Swap, R., Lefer, B., Valin, L., Lee, S.C., Fioletov, V., Zhao, X., Davies, J., Williams, D., Abuhassan, N., Shalaby.: Pandora: Connecting in-situ and satellite monitoring in support of the Canada–US Air Quality Agreement. EM: Air Waste Manag. Assoc. Mag. Environ. Managers, 2470–4741 (2019). https://www.awma.org/empastissues

US Environmental Protection Agency (US-EPA).: Integrated science assessment for oxides of nitrogen—health criteria. EPA/600/R-15/068. US Environmental Protection Agency, Research Triangle Park (2016). https://www.epa.gov/isa/integrated-science-assessment-isa-nitrogen-dioxide-health-criteria

van Donkelaar, A., Martin, R.V., Brauer, M., Hsu, N.C., Kahn, R.A., Levy, R.C., Lyapustin, A., Sayer, A.M., Winker, D.M.: Global estimates of fine particulate matter using a combined geophysical-statistical method with information from satellites, models, and monitors. Environ. Sci. Technol. **50**(7), 3762–3772 (2016). https://doi.org/10.1021/acs.est.5b05833

World Bank (2022). The Global Health Cost of PM$_{2.5}$ Air Pollution: A case for action beyond 2021. International Development in Focus. World Bank, Washington DC. https://openknowledge.worldbank.org/handle/10986/36501

World Health Organization (WHO).: WHO ambient air quality database 2022. WHO, Geneva, Switzerland (2022). https://www.who.int/data/gho/data/themes/air-pollution/who-air-quality-database

Chapter 5
Ambient Air Quality Monitoring

Abstract Timely, reliable, and accurate data on the ambient concentrations of key pollutants is critical for making informed decisions about public health protection, evaluating the effectiveness of control strategies, and tracking progress toward meeting air quality goals, as well as informing citizens about the quality of the air they breathe and, in the event of severe air pollution episodes, implementing additional protection and control measures. In this context, this chapter reviews the state of air quality monitoring in each Southeast Asian country, covering air quality monitoring networks operated for regulatory and warning purposes, networks of monitors assessing directly or indirectly specific air quality's aspects, and the use of low-cost monitors to complement existing air quality monitoring systems. The pollutant species monitored and the concentration thresholds set as air quality standards in each country are examined, as well as the air quality indices used to facilitate the dissemination and understanding of air quality conditions. The reader will also learn about international initiatives, non-profit organizations, and private groups that compile and archive ground-level air quality monitoring data and make it available to the public.

Keywords Criteria pollutants · Air quality standards · Air quality index · Acid deposition · Low-cost sensors

The ambient concentrations of pollutants used to prepare the State of Global Air report and products derived from it are obtained from ground-level monitoring sites installed at strategic locations, most often by governmental authorities, in accordance with guidelines for the design and operation of air quality monitoring networks, which include protocols for quality system requirements, approved monitors and methodologies, measurement quality checks, instrument calibration and maintenance, and data quality assurance, reporting, and archiving (e.g., US-EPA 2016). The data collected at each monitoring station is usually aggregated hourly via telemetry by the responsible agency.

Pollutant species measured by regulatory air quality monitoring networks are typically limited to what are known as "criteria pollutants." These are regulated

E. Velasco et al., *Air Quality Management and Research in Southeast Asia*,
SpringerBriefs in Earth System Sciences, https://doi.org/10.1007/978-3-031-69088-4_5

pollutants, whose permissible concentrations are based on human health and environmental criteria in accordance with science-based guidelines (e.g., World Health Organization (WHO) air quality guidelines; WHO 2022a). Criteria pollutants can differ by country, but they are typically six: carbon monoxide (CO), nitrogen dioxide (NO_2), sulfur dioxide (SO_2), ozone (O_3), lead (Pb), and particulate matter including one or more of the following fractions: total suspended particles (TSP), particles smaller or equal to 10 and 2.5 μm in size (PM_{10} and $PM_{2.5}$, respectively). The concentration readings of these pollutants are used to calculate different interval (annual, 1-, 8- or 24-h) average concentrations in accordance with standards and regulations established in each country to protect public health, and inform about the quality of the air; and in cases of severe pollution, so that authorities can implement opportune protection and control measures. Figure 5.1 shows the deployment of an air quality monitoring station equipped with regulatory-grade monitors.

Table 5.1 shows the concentration thresholds set as air quality standards by each Southeast Asian country. Brunei Darussalam, Myanmar, Timor-Leste, and Vietnam have yet to develop their own standards. Compared to the WHO's air quality guideline (AQG) levels and interim targets, the current air quality standards in Southeast Asian countries fall between interim targets #1 and #3 (interim targets #1 and #4 are the first and last sets of thresholds recommended before achieving the AQG levels, respectively). Only Indonesia, Malaysia, the Philippines, and Singapore have O_3 standards that are equal to or stricter than the WHO's AQG levels. In the case of $PM_{2.5}$, Indonesia, Lao PRD and Singapore have standards that are higher than interim target #4 but equal to or lower than target #3. It should be noted that Singapore does not have air quality standards, but rather air quality targets. Indonesia and the Philippines air quality standards for PM_{10} and NO_2 are still above interim target #1. The same is true for these two countries' SO_2 standards, as well as those of Cambodia, Lao PDR, and Thailand.

Environmental agencies use indices to relate the concentration of each criteria pollutant to an air quality category and thus assign a level of health risk based on the highest value found, to make air quality information more widely available and understandable. Depending on the country, they can take different names: Air Quality Index, Air Pollution Index, Pollutant Standard Index, or another similar. All of these indices, however, are similar to that developed by the US-EPA (https://www.airnow.gov/aqi/aqi-basics/). These indices are typically published on official websites and mobile applications, and in some cases, alongside the measured concentrations of each pollutant species.

Table 5.2 lists the criteria pollutants measured in each country of Southeast Asia, as well as the environmental agencies in charge of measuring them. The selection of these pollutants is dictated by each country's legislation, though other pollutant species are occasionally measured due to their toxicity or potential to form O_3 and secondary aerosols, as is the case with certain volatile organic compounds (VOCs) in Singapore and Malaysia. The information presented here was obtained by reviewing the official websites of national environmental authorities, as well as international assessments on the subject (e.g., United Nations Environment Programme (UNEP) individual reports on national air quality policies, UNEP 2021a).

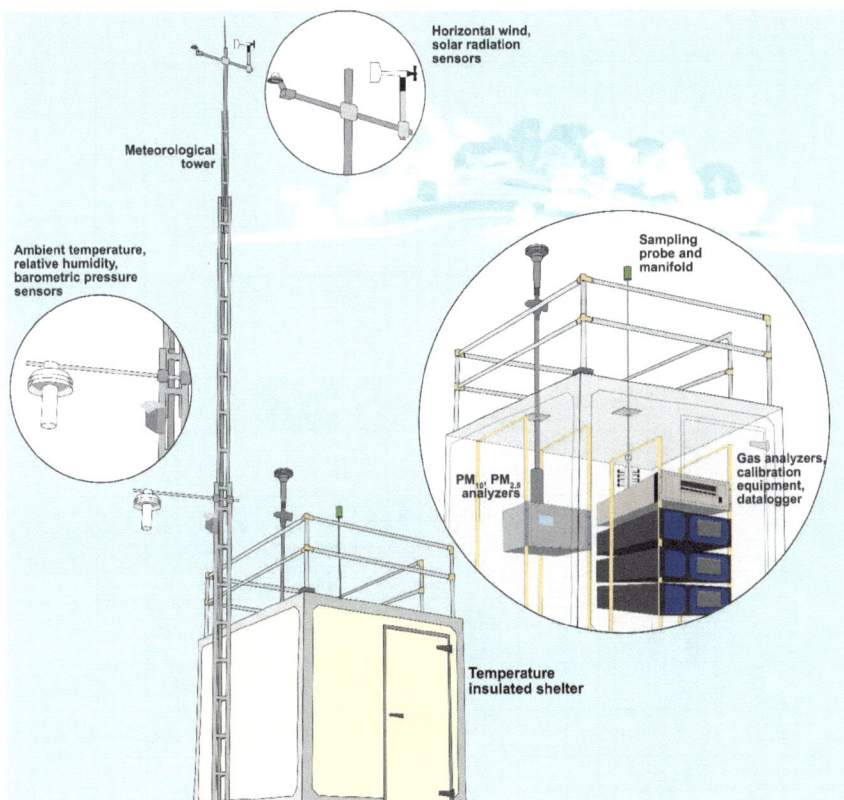

Fig. 5.1 For regulatory purposes, air quality monitoring stations, such as the one shown here, are equipped with reference- or equivalent-grade monitors for measuring criteria pollutants. Monitoring stations are also equipped with instruments for measuring meteorological variables, including ambient temperature (T), relative humidity (RH), atmospheric pressure (P), wind speed (w_s), wind direction (w_d), incoming shortwave radiation ($K_\downarrow$), and precipitation. Analytical monitors must be placed indoors under controlled conditions; shelters designed specifically for this purpose are often used. Air samples must be taken from heights above the urban canopy, avoiding any obstacles or structures that may disturb the wind flow. Teflon sampling lines mounted on lattice towers can be used for this purpose

A few Southeast Asian countries still lack regulatory air quality monitoring programs. Timor-Leste lacks an air quality monitoring station. Lao PDR and Myanmar do not appear to have any operational monitoring stations at this time; if they do, they are not publicly reported. Malaysia, the Philippines, Singapore, and Thailand are on the opposite side. These four countries have well-established air quality monitoring networks. The networks in the Philippines, Singapore, and Thailand stand out for having user-friendly web portals that provide real-time information to the public, including both air quality indices and pollutant concentrations measured at each station. In the middle are Indonesia and Vietnam, which are building their own

Table 5.1 Air quality standards set by each Southeast Asian nation, along with the WHO's recommended air quality guideline (AQG) levels and interim targets (IT). A color gradient is used to highlight the WHO's recommended AQG levels and IT, as well as in the air quality standards of each country based on whether or not they are equal to or higher than them

Country	Standard /guideline /target	PM$_{2.5}$ ($\mu g\,m^{-3}$) Annual	24-h[a]	PM$_{10}$ ($\mu g\,m^{-3}$) Annual	24-h[a]	TSP ($\mu g\,m^{-3}$) Annual	24-h[a]	O$_3$ ($\mu g\,m^{-3}$) Annual/Peak season[b]	8-h[a]	1-h	NO$_2$ ($\mu g\,m^{-3}$) Annual	24-h[a]	1-h	SO$_2$ ($\mu g\,m^{-3}$) Annual	24-h[a]	1-h	CO ($mg\,m^{-3}$) 24-h[a]	8-h	1-h	Lead ($\mu g\,m^{-3}$) Annual	24-h[a]
WHO	IT-1	35	75	70	150	--	--	100[b]	160	--	40	120	--	--	125	--	7	--	--	--	--
	IT-2	25	50	50	100	--	--	70[b]	120	--	30	50	--	--	50	--	--	--	--	--	--
	IT-3	15	37.5	30	75	--	--	--	--	--	20	20	--	--	--	--	--	--	--	--	--
	IT-4	10	25	20	50	--	--	--	--	--	--	--	--	--	--	--	--	--	--	--	--
	AQG level	5	15	15	45	--	--	60[b]	100	--	10	25	--	--	40	--	4	--	--	--	--
Brunei Darussalam[c]	--	--	--	--	--	--	--	--	--	--	--	--	--	--	--	--	--	--	--	--	--
Cambodia[d]	National	25	50	50	120	100	330	--	--	200	--	100	300	100	300	500	--	20	40	--	5
Indonesia[e]	National	15	65	150	--	--	--	50	--	235	100	--	400	60	260	400	10	--	30	--	--
	Jakarta	15	65	150	--	--	--	30	--	200	60	--	400	60	365	400	9	--	26	--	--
Lao PDR[f]	National	15	50	50	120	100	330	--	140	200	38	--	207	--	131	341	--	10.4	35	0.15	--
Malaysia[g]	National IT-1	35	75	50	150	--	--	--	120	200	--	75	320	--	105	350	--	10	35	--	--
	National IT-2	25	50	45	120	--	--	--	120	200	--	75	300	--	90	300	--	10	35	--	--
	National Standard	15	35	40	100	--	--	--	100	180	--	70	280	--	80	250	--	10	30	--	--
Myanmar[h]	National	--	--	--	--	--	--	--	--	--	--	--	--	--	--	--	--	--	--	--	--
Philippines[i]	National Air Quality Guideline Values	25	50	150	60	90	230	--	60	140	--	150	--	80	180	--	--	10	35	1	1.5
Singapore[j]	Targets by 2020	12	37.5	20	50	--	--	--	100	--	40	--	200	--	50	200	--	10	35	--	--
	Long term targets	10	25	--	--	--	--	--	--	--	--	--	--	--	20	--	--	10	30	--	--
Thailand[k]	National	15	37.5	50	120	100	330	--	140	200	57	--	320	100	300	780	--	10.2 6	34.2	1.5	--
Timor-Leste[l]	--	--	--	--	--	--	--	--	--	--	--	--	--	--	--	--	--	--	--	--	--
Vietnam[m]	--	--	--	--	--	--	--	--	--	--	--	--	--	--	--	--	--	--	--	--	--

Table 5.1 (continued)

[a] 99th percentile (i.e., 3–4 exceedance days per year)

[b] Average of daily maximum 8-h mean O_3 concentration in the six consecutive months with the highest six-month running-average O_3 concentration

[c] Brunei Darussalam does not have air quality standards at this time

[d] Clean Air Plan of Cambodia (Cambodia Ministry of Environment 2021)

[e] National air quality standards: Indonesian Government regulation PP No. 41/1999 (https://www.menlhk.go.id/). Jakarta's air quality standards: Governor's Decree DKI Jakarta No. 55/2001 (DKI and Vital Strategies 2019)

[f] NO_2, SO_2, and CO standards are converted from ppm to $\mu g\ m^{-3}$ and mg m^{-3} as corresponded considering standard ambient conditions of 25 °C and 1013 mb (Lao PDR Ministry of Natural Resources and the Environment 2017). Agreement on National Environmental Standards No. 2734/PMO.WREA (Prime Minister's Office 2009). Lead standard is based on a monthly base

[g] Malaysia has established two interim targets and air quality standards that should have been met in 2015, 2018 and 2020, respectively (Malaysia's Department of Environment 2022)

[h] An Environmental Conservation Law has been promulgated, but air quality regulations and standards have not been developed (World Bank 2019)

[I] National Air Quality Report 2016–2018 (Department of Environment and Natural Resources 2020)

[j] National Environmental Agency—Air Quality (https://www.nea.gov.sg/our-services/pollution-control/air-pollution/air-quality)

[k] Thailand also counts with air quality standards for nine toxic pollutants: carbon disulfide, benzene, vinyl chloride, 1,2-dichloroethane, trichloroethylene, dichloromethane, 1,2-dichloropropane, tetrachloroethylene, chloroform, and 1,3-butadiene (see Air4Thai). The air quality standard for Pb is based on a monthly base (https://air4thai.pcd.go.th/webV2/)

[l] No information available. It does not appear that air quality standards have been enacted in Timor-Leste at this time

[m] No information available. It does not appear that air quality standards have been enacted in Vietnam at this time

air quality monitoring systems. Brunei Darussalam does have a monitoring system in place, but there is little information about it available to the public.

At the time of writing, 390 air quality monitoring stations in Southeast Asia were in operation, including five stations measuring $PM_{2.5}$ in US embassies and consulates through the AirNow-DOS program. This number of monitoring stations is clearly insufficient to assess the level of air pollution to which Southeast Asian residents are exposed, as well as to ensure proper air quality management. There are 0.6 monitoring stations per million inhabitants across the region, but the density of stations per country ranges from zero to sixteen (see Fig. 5.2). Citizens from Lao PDR, Myanmar, and Timor-Lester, totaling 63.5 million, have essentially no access to air quality data, whereas the density of monitoring stations in Brunei Darussalam and Singapore is similar to or higher than in countries in Europe and North America (≥ 4 stations per 1-million inhabitants, Martin et al. 2019).

The following sections of this chapter provide an overview of the status of each country's monitoring system, international and regional monitoring projects covering Southeast Asia, and of global initiatives focused to compile available air quality data. The chapter closes by discussing the use of low-cost monitors to complement existing air quality monitoring networks.

Table 5.2 Official sources of information on the ambient air quality monitoring systems in each Southeast Asian country and the criteria pollutants they measure. Only monitoring stations equipped with regulatory-grade instrumentation are taken into account. Question marks indicate that no information was found or that the available information is uncertain

Country	Government Department/Agency	Pollutants								Number of monitoring stations	Start of monitoring	Data repository and public posting
		CO	NO$_2$	SO$_2$	O$_3$	TSP	PM$_{10}$	PM$_{2.5}$	Pb			
Brunei Darussalam[a]	Department of Environment, Parks, and Recreation (DEPR), Ministry of Development	–	–	–	–	–	✓	✓	–	7	2015 (?)	DEPR website: http://www.env.gov.bn/Theme/Home.aspx
Cambodia[b]	Ministry of Environment	✓	✓	✓	✓	✓	✓	✓	–	10	1999	Ministry of Environment and Forestry
Indonesia[c]	Meteorology, Climatology, and Geophysics Agency (BMKG), Ministry of Environment and Forestry (KLHK), Environmental Agency of Jakarta (DKI Jakarta)	✓	✓	✓	✓	✓	✓	✓	–	BMKG: 24, KLHK: 46, DKI Jakarta: 5[1]	1998	BMKG website and mobile app.: https://www.bmkg.go.id/ KLHK website and IQAir platform: https://ispu.menlhk.go.id/ DLH Jakarta website and Jaki—Jakarta Kini mobile app: https://lingkunganhidup.jakarta.go.id/
Lao PDR[d]	Ministry of Natural Resources and the Environment	–	✓	✓	–	✓	✓	–	–	1[1] (?)	2002 (?)	(?)

(continued)

Table 5.2 (continued)

Country	Government Department/Agency	Pollutants								Number of monitoring stations	Start of monitoring	Data repository and public posting
		CO	NO$_2$	SO$_2$	O$_3$	TSP	PM$_{10}$	PM$_{2.5}$	Pb			
Malaysia[e]	Department of Environment, Ministry of Environment and Water (DOE-ME&W)	✓	✓	✓	✓	–	✓	✓	✓	65	1979	DOE-ME&W's website and MyIPU mobile app.: http://apims.doe.gov.my/home.html
Myanmar[f]	Department of Meteorology and Hydrology	–	–	–	–	–	–	–	–	1¹ (?)	–	(?)
Philippines[g]	Air Quality Management Section (AQMS), Environmental Management Bureau	✓	✓	✓	✓	✓	✓	✓	✓	98	1975	AQMS website and Philippines AQI mobile app.: https://air.emb.gov.ph/
Singapore[h]	National Environmental Agency (NEA)	✓	✓	✓	✓	–	✓	✓	–	22	1970	NEA website and myENV app.: https://www.haze.gov.sg/
Thailand[i]	Thailand Pollution Control Department	✓	✓	✓	✓	✓	✓	✓	✓	96	1983	Air4Thai website and mobile application: https://air4thai.pcd.go.th/webV2/
Timor-Leste[j]	Secretary State for Environment	–	–	–	–	–	–	–	–	0	–	–

(continued)

Table 5.2 (continued)

Country	Government Department/Agency	Pollutants								Number of monitoring stations	Start of monitoring	Data repository and public posting
		CO	NO$_2$	SO$_2$	O$_3$	TSP	PM$_{10}$	PM$_{2.5}$	Pb			
Vietnam[k]	General Department of Environment	✓	✓	✓	✓	–	✓	✓	–	45[l]	2009	Department of Environment website and Envisoft mobile app.: http://enviinfo.cem.gov.vn/

[a] Air quality management in Brunei Darussalam (Department of Environment, Parks, and Recreation 2021)

[b] Clean Air Plan of Cambodia (Cambodia Ministry of Environment 2021). It is not clear if all criteria pollutants are measured at all stations. Most stations are installed in Phnom Penh

[c] In Indonesia, air quality monitoring is carried out by both national agencies and local governments

[d] No updated information is available on air quality monitoring in Lao PDR (World Bank 2006; Lao PDR Ministry of Natural Resources and the Environment 2012)

[e] 2021 Environmental Quality Report (Malaysia's Department of Environment 2022)

[f] Two air quality monitoring stations, one in Yangon and other in Mandalay, are apparently in operation (Myint 2021; World Bank 2019)

[g] National Air Quality Report 2016–2018. (Department of Environment and Natural Resources 2020)

[h] Environmental Protection Division Annual Report 2018 (National Environmental Agency 2019)

[i] Pollution Control Department. The official website was migrating to a new platform at the time of writing (https://air4thai.pcd.go.th/)

[j] No information available. Apparently Timor-Leste does not have monitoring stations

[k] Department of Environment website (http://enviinfo.cem.gov.vn/)

[l] Including air quality monitoring stations of the US Department of State's AirNow program: 2 in Indonesia, 1 in Lao PDR, 1 in Myanmar, and 2 in Vietnam

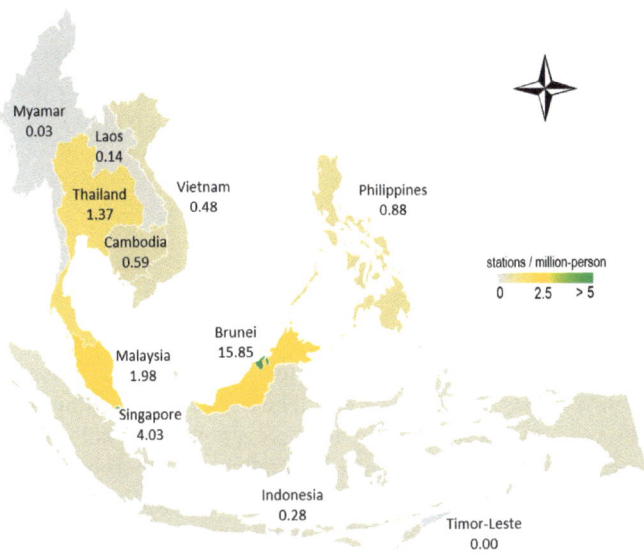

Fig. 5.2 Number of air quality monitoring stations per million inhabitants in Southeast Asian countries based on data from 2022

5.1 Air Quality Monitoring Status in Each Country

This section provides an overview of the status of air quality monitoring in each Southeast Asian country. The progress achieved in each country is pointed out, as well as the main shortcomings and challenges that each one faces in terms of instrumentation and environmental regulations. Table 5.2 lists the criteria pollutants monitored in each country, as well as the number of operational monitoring stations at the time of writing and the year when air quality monitoring began. The corresponding references to official institutions in charge of air quality monitoring, as well as links to their internet portals where the public can access air quality data, are also included. The information presented here can be complemented with the revision that Oanh et al. (2023) did on the air quality status in Myanmar, Thailand, and Vietnam, and a similar revision that Cambaliza et al. (2023) did focusing on the cities of Manila, Kuala Lumpur, Singapore, and Jakarta.

5.1.1 Brunei Darussalam

The Department of Environment, Parks, and Recreation operates seven monitoring stations spread across the Sultanate's four districts (Department of Environment, Parks, and Recreation 2021). PM_{10} and $PM_{2.5}$ are the pollutants being monitored. Their readings are used to compute a Pollutant Standard Index (PSI), which is published on the Department's website every day. Both particle fraction concentrations are not published. Only PSI records computed on the same day are made public; historical records are not. Similarly, neither the location of the monitoring stations nor the characteristics of the monitors, nor the algorithm used to calculate the 24-h PSI, are available to the public.

5.1.2 Cambodia

Cambodia published its first plan to achieve clean air in November 2021 (Cambodia Ministry of Environment 2021). This plan outlines the current state of air quality in the country and the policies and actions that will be implemented to improve air quality management. One of these actions is to improve air quality monitoring capabilities. Across the country, there are currently ten operational monitoring stations equipped with regulatory-grade monitors. To date, 44 stations equipped with low-cost monitors have been installed to extend air quality monitoring coverage to more provinces, at the expense of collecting unreliable data. The daily averages of the pollutants measured at each monitoring station, as well as an Air Quality Index (AQI) calculated by themselves, are published on the Ministry of the Environment's Facebook page. Historical records appear to be unavailable to the public.

5.1.3 Indonesia

Air quality monitoring in Indonesia is carried out at the national level by two agencies, as well as at the regional or local level by individual provinces. The Ministry of Environment and Forestry maintains 46 monitoring stations across the country, while the Agency for Meteorology, Climatology, and Geophysics maintains 24 monitoring stations. The Environment Agency of the Special Capital Region (*Daerah Khusus Ibukota*, DKI) of Jakarta operates five fixed stations and three mobile stations on a regional scale (DKI and Vital Strategies 2019). Air quality data collected by the Ministry of Environment and Forestry is published in real time on the ministry's website. Likewise, the other two agencies publish the air quality data they collect on their own websites and through mobile applications designed for that purpose. The ambient concentrations of PM_{10}, NO_2, SO_2, CO, and O_3 are used to calculate an Air Pollutant Standard Index (APSI) that is released at least twice daily, whereas $PM_{2.5}$ concentrations are used to calculate a second index that is released every hour. It appears that APSI's historical records are accessible to the public, but it is unclear whether the same is true for pollutant concentrations.

To improve air quality assessments across the country, the Ministry of Environment and Forestry deploys more than 1,700 passive samplers for measuring NO_2 and SO_2 across all Indonesian provinces each year to complement data from automatic monitoring stations. Nitrogen dioxide is used to trace pollution from vehicular traffic, while SO_2 is used to trace pollution from industries and diesel vehicles. These two pollutants are used to calculate a provincial Air Quality Index (Ministry of Environment and Forestry 2020).

5.1.4 Lao PDR

In Lao PDR, there is no updated information on air quality monitoring. The information available ranges from 2006 to 2012 (World Bank 2006; Lao PDR Ministry of Natural Resources and Environment 2012). Although air quality standards have been established, there is no information available to confirm the presence of operational monitoring stations at the time of writing.

5.1.5 Malaysia

Malaysia's ambient air quality is continuously monitored by a network of 65 monitoring stations. They are strategically located in urban (11), suburban (4), industrial (7), and rural (12) regions (Malaysia's Department of Environment 2022). An additional station, located in central Peninsular Malaysia at Jerantut, measures regional background concentrations (Latif et al. 2014). Selected stations are equipped with instruments for measuring non-methane hydrocarbons (NMHCs). In addition, nineteen stations are equipped with high volume samplers to collect PM_{10} samples once every six days for analysis of heavy metals, such as lead, mercury, iron, and copper. The Air Pollution Index, calculated from measured criteria pollutants, is made available to the public on an hourly basis via a website and a mobile application run by Malaysia's Department of Environment. Hourly indices for the last seven days are available for download, but not for previous dates. Similarly, it appears that concentrations for each criteria pollutant are not publicly available.

5.1.6 Myanmar

Myanmar has no air quality regulations and any monitoring system in operation. However, two monitoring stations, one in Yangon and one in Mandalay, appear to be operational. They are operated by Myanmar's Department of Meteorology and Hydrology with Japanese assistance, but the data is not made public (Myint 2021; World Bank 2019). Similarly, the Heinrich Böll Foundation operates another monitoring station in a residential district of Yangon, and the data is used to calculate an Air Quality Index, which is available upon request (https://www.boell.de/en). There are plans to install seven air quality monitoring stations, three in Yangon, three in Mandalay, and one in Nay Pyi Taw, as well as to instrument five mobile stations to cover other regions of the country. Both fixed and mobile stations were expected to be in operation by 2023 at the time of writing (Myint 2021).

5.1.7 Philippines

There are currently 32 air quality monitoring stations in Metro Manila and 66 in the rest of the Philippine archipelago. The Air Quality Index, calculated from the ambient concentrations of the criteria pollutants measured at each monitoring station, is published in real-time. Historical data on the annual average concentrations of each pollutant is also publicly available. It is worth noting that both the legislation and technical manuals that regulate local air quality management are made available to the public. Air quality data is easily accessible on the website of the Environmental Management Bureau's Air Quality Management Section.

5.1.8 Singapore

Singapore's air quality monitoring network includes 22 fixed stations, 18 measure ambient air quality and four roadside air pollution. The latter are used to evaluate the effectiveness of emission control programs for vehicles. In addition to measuring criteria pollutants, selected VOCs are also measured. It has a manual air sampling program that monitors Pb levels in particulate matter every two weeks and a program that monitors dioxin levels every year. Except for $PM_{2.5}$, ambient concentrations of measured criteria pollutants are used to calculate a local Pollutant Standard Index to inform citizens about air quality conditions. Along with this index, the hourly moving average concentrations of each pollutant and the 1-h $PM_{2.5}$ concentrations are made available to the public. These data are grouped into five regions, which correspond to how the island city-state is divided for air quality purposes. Hourly $PM_{2.5}$ concentrations are provided as an indication of the current air quality conditions through a system of four descriptor bands that act as a guide for the public to adjust immediate activities. Pollutant concentrations are only made available to the public on a day-to-day basis. The following day, hourly data from the previous day are restricted. Only historical PSI records from 2009 are available to the public. Molina et al. (2019) provide a comprehensive description of Singapore's air quality monitoring system.

5.1.9 Thailand

Thailand has 96 air quality monitoring stations and 10 mobile stations spread across the five regions into which the country is divided for air quality management purposes. Thirty-two stations are located in Bangkok and surrounding provinces. Twenty-five provinces have at least one monitoring station. They are operated by the Pollution Control Department. Every hour, the moving average concentrations of six criteria pollutants, as well as the Air Quality Index calculated from them are published for each monitoring station. These data are displayed using maps, graphs, and tables. Historical records can be downloaded from the Air4Thai website (https://air4thai.pcd.go.th/).

5.1.10 Timor-Leste

Timor-Leste has not enacted air quality legislation and thus has not implemented a monitoring system. In October 2021, the UNEP approved a project to improve early warning systems for increased climate resilience, which appears to include an air quality monitoring system to mitigate the effects of climate change and air pollution impacts (UNEP 2021a,b).

5.1.11 Vietnam

Vietnam has yet to enact air quality regulations. Nonetheless, an air quality monitoring network is already in place. It consists of 45 monitoring stations, the majority of which are located in the metropolitan areas of Hanoi and Ho Chi Minh City, as well as surrounding provinces. The authorities use the US-EPA's Air Quality Index to provide real-time information on the status of air quality. The Department of Environment makes this index and the associated concentrations of criteria pollutants available to the public through a website and a mobile application. It is unclear whether historical records are open to the public. The National Plan on Air Quality Management 2021–2025 (Decision 1973/qD-TTg) was promulgated in November 2021 (Vietnam Environment Administration 2021). It includes directives to develop air quality standards and strengthen air quality monitoring. The former were scheduled to go into effect in 2023.

5.2 International and Regional Air Quality Monitoring Initiatives

There are four international initiatives assessing directly or indirectly certain aspects of air quality worldwide, including monitoring sites in Southeast Asia. This section provides a brief overview of them, as well as the Acid Deposition Monitoring Network in East Asia, which also includes monitoring sites throughout the region. Table 4.1 summarizes the details of these initiatives.

5.2.1 AirNow-DOS Program

The US Department of State's AirNow program (AirNow-DOS; https://www.airnow. gov/) is another source of air quality data. AirNow-DOS collects air quality monitoring data from US embassies and consulates around the world. The program has six regulatory-grade $PM_{2.5}$ monitors in Southeast Asia, two in Jakarta and one each in Hanoi, Ho Chi Ming City, Yangon, and Vientiane. Both real-time and historical data are available to the public. Dhammapala (2019) provides a detailed description of this program as well as an analysis of the $PM_{2.5}$ data collected worldwide.

5.2.2 AEronet RObotic NETwork (AERONET)

The AEronet RObotic NETwork (AERONET) is a global network that uses ground-based sun photometers to measure aerosol optical properties as a proxy for aerosol abundance (Holben et al. 1998; https://aeronet.gsfc.nasa.gov/). Sun photometers are passive remote-sensing instruments that measure the sun's direct light. The measured light is not equal to that emitted by the sun because absorption and scattering by particles, gases, and water vapor reduce the solar flux. After accounting for light attenuation due to Rayleigh scatter, absorption by O_3 and other pollutants gases, water vapor abundance, and ruling out cloud interference, the attenuation of sunlight due to particles along a column from the surface to the top of the atmosphere, known as aerosol optical depth (AOD), is determined by applying processing algorithms. AOD is a unitless measure based on the Beer-Lambert-Bouguer law, according to which it can have values of 0.05 or less on a very clear day, but values greater than 1.0 under extremely hazy conditions. AOD depends on the particle properties, including mass, size, chemical composition, optical properties, and vertical distribution. Sun photometers also provide information on the amount of water vapor in the atmosphere (precipitable water) and the aerosol size distribution using the Ånsgtröm parameter relationship.

The US National Aeronautics and Space Administration (NASA) established AERONET in the mid-1990s, and it has since expanded significantly, with more than 600 sun photometers spread across the globe, including 25 in Southeast Asia. The network's success has been based on the standardization of instruments, calibration, data processing, and the free distribution of AOD data and related aerosols databases. The AERONET website offers AOD data with three quality levels: Level 1.0 (unscreened), Level 1.5 (cloud-screened and quality controlled), and Level 2.0 (quality assured). Processing algorithms are constantly evolving in order to reduce uncertainties, improve accuracy, and make processed data available faster. Currently, after post-field instrument calibration, Level 2.0 quality-controlled data can be obtained within a month (Giles et al. 2019).

5.2.3 Pandonia Global Network

NASA and the European Space Agency (ESA) launched the Pandonia Global Network in 2019 to establish a global network of ground-based standardized sun photometers for measuring total column and vertically resolved concentrations of a suite of pollutant species, with a focus on O_3, NO_2, and formaldehyde (HCHO) (Szykman et al. 2019; https://www.pandonia-global-network.org/). The network provides real-time, standardized, calibrated, and verified air quality data and associated uncertainties values to support the validation and verification of satellite UV–visible sensors (see Chapter 8). It has been widely used to validate satellite observations as well as to support a number of field campaigns studying the chemical composition of the atmosphere. It is expected to improve existing ground-level air quality monitoring networks and help to better understand emissions, chemistry, and meteorology of airborne pollutants as part of future air quality systems based on satellite observations. At the time of writing, nine measurement sites were available for Southeast Asia; more sun photometers are expected to be installed in the region in the near future.

5.2.4 SPARTAN Network

Surface Particulate Matter Network (SPARTAN) is a grass-roots network run by individual scientists. The network makes available to the public data on $PM_{2.5}$ mass concentration, chemical composition, and optical characteristics for air quality management and satellite remote-sensing retrieval (Snider et al. 2015, 2016; https://www.spartan-network.org/). It started in 2012 and has grown to 29 active sites, including four sites in Southeast Asia: Bandung, Manila, Singapore, and Hanoi. To measure AOD, measurement sites are intentionally collocated with AERONET Sun photometers. The network uses high-degree but not regulatory-grade instrumentation. A nephelometer is used to measure backscatter and total scatter. Particle samples collected on filters for nine days are chemically analyzed offline, including the determination of water-soluble ions, equivalent black carbon (eBC), and trace elements.

The organic components are derived from the difference between the total particle mass and the mass of the analyzed species. This approach has the drawback of not accounting for semi-volatile material, which can represent a significant fraction of the particle load (Moya et al. 2011). Time resolved $PM_{2.5}$ mass concentrations are obtained by combining filter-based $PM_{2.5}$ mass and total scatter to calculate $PM_{2.5}$ mass at finer temporal resolution than filters alone.

5.2.5   *Acid Deposition Monitoring Network in East Asia (EANET)*

The Acid Deposition Monitoring Network in East Asia (EANET, https://www.eanet. asia/) was established in 2001 as an intergovernmental initiative to create a common understanding on the state of acid deposition problems in East Asia, with the aim of providing information for decision-making and promoting regional cooperation (Akimoto et al. 2022). EANET currently has thirteen countries participating, eight from Southeast Asia. EANET manages three types of monitoring stations: urban, rural, and remote. Some of these stations have ecological survey sites attached to collect data for assessing the effects of acidification on terrestrial ecosystems in soil and vegetation, as well as inland aquatic environments. Figure 5.3 shows the locations of the 33 acid deposition sites in Southeast Asia. All participants adhere to EANET-established common methodologies for wet and dry deposition sampling, analysis, and data reporting (EANET 2022). On a daily or weekly basis, wet deposition samples are analyzed for pH, electric conductivity, anions (SO_4^{2-}, NO_3^-, Cl^-, F^-, PO_4^{3-}), cations (Na^+, K^+, NH_4^+, Ca^{2+}, Mg^{2+}, HCO_3^+), and organic acids. Dry deposition includes measurements of anions and cations in particulate matter in addition to measurements of gaseous species (SO_2, HNO_3, HCl, NH_3) collected by filter packs and passive samplers. Some sites are equipped with automatic air quality monitors for measuring SO_2, O_3, NO, NO_2, PM_{10}, and $PM_{2.5}$.

5.3   Compilations of Ambient Air Quality Monitoring Data

The WHO's Air Quality Database and the Tropospheric Ozone Assessment Report (TOAR) are international initiatives that compile and archive all air quality data available from ground-level air quality monitoring stations around the world. Although their databases are not up to date, and they may not include data from all existing monitoring stations, they are the way to access air quality data without going through the environmental agencies of each country. Non-profit organizations such as OpenAQ, as well as a few private groups, are also working to compile and make publicly available air quality databases.

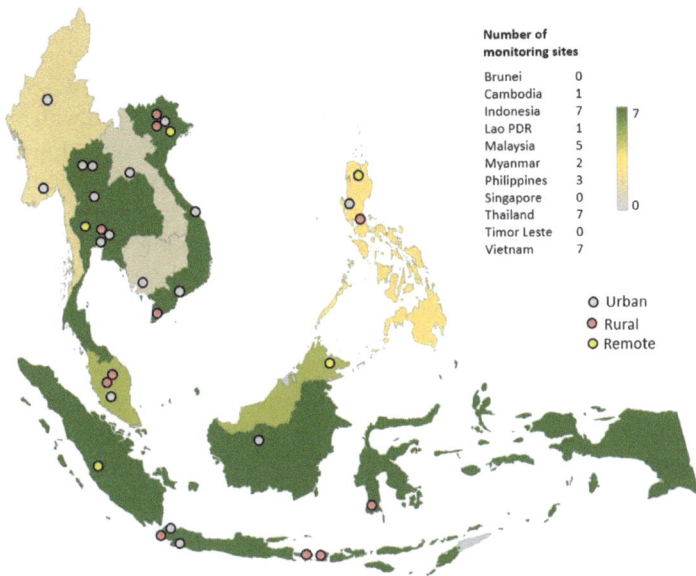

Fig. 5.3 Acid Deposition Monitoring Network in East Asia (EANET) monitoring sites in Southeast Asia as of 2021 (EANET, 2022). Urban sites can be found in urban, suburban, and industrial areas. Rural sites are located in forests and agricultural fields at least 20 km away from major emission sources such as cities, power plants, and highways. Remote sites must be representative of long-range transport deposition and must be at least 50 km away from any large emission source, and 500 m away from major roads

The WHO's Ambient Air Quality Database compiles ground-level measurements of annual mean concentrations of NO_2, PM_{10}, and $PM_{2.5}$ on an urban scale rather than at individual monitoring stations (WHO 2022b; https://www.who.int/data/gho/data/themes/air-pollution). It collects data representative of pollutant concentrations at ambient level from national and subnational air quality monitoring systems, as well as regional and global programs such as AirNow-DOS. Data from monitoring stations measuring regional background pollution, as well as data from stations monitoring "hot spots" of air pollution, such as highways with exceptionally heavy traffic and areas impacted by a specific industry, are not included. When available, the database includes references to data sources as well as the temporal coverage (number of days per years covered by measurements). It is updated every two to three years, with the most recent update occurring in 2022, in which measurements from 2010 to 2019 were covered. It includes data from 398 urban settlements in 106 cities in all Southeast Asian countries with the exception of Brunei Darussalam, Cambodia, and Timor-Leste.

The Tropospheric Ozone Assessment Report (https://toar-data.org) compiles and makes publicly available a suite of O_3 data products including standard statistics, health and vegetation impact metrics, and trend information from data obtained from thousands of measurement sites around the world, with the mission of assessing the global distribution and trends of tropospheric O_3. However, original hourly time series are not freely available (Schultz et al. 2017). TOAR is a scientific initiative supported by the International Global Atmospheric Chemistry (IGAC) project. The TOAR database only stores data from monitoring stations equipped with regulatory-grade instrumentation and measurement periods exceeding two years. Environmental agencies and research organizations provide the data. Users can extract data for a specific region, country, or measurement site through a web interface that unfortunately requires database management knowledge. The majority of available data sets are centered in Europe, North America, and East Asia. There are only 13 datasets available for Southeast Asia: 5, 4, 3, and 1 from Indonesia, Thailand, Malaysia, and Vietnam, respectively.

OpenAQ is a community-driven platform that compiles and archives air quality data from around the world with the goal of empowering the public to fight air pollution using open-source tools and cooperation (https://openaq.org/). Air quality data is compiled from real-time public sources, typically provided by environmental agencies and international organizations. It includes ambient concentrations of PM_{10}, $PM_{2.5}$, O_3, SO_2, CO, NO_2, and black carbon from individual monitoring stations equipped with regulatory-grade instrumentation. The web interface is user-friendly, but the public should be aware that real-time data are preliminary and may have uncertainties as they have not been fully validated by the air quality monitoring networks themselves; this is especially true for air quality programs with inadequate quality assurance and quality control (QA/QC) schemes. At the time of writing, OpenAQ was testing a pilot platform that would include data from low-cost monitors. Indeed, these monitors allow for the filling of data gaps in air quality in many places, particularly in developing countries, but users must be aware of the poor reliability of the data they provide. OpenAq does not guarantee data quality, so data compiled from low-cost monitors must not be used for regulatory and warning purposes and must be used with extreme caution in academic research. At the time of writing, 215 of the 288 data sets compiled by OpenAQ for Southeast Asia came from low-cost monitors. No data were available from Brunei Darussalam and Cambodia, and the number of monitoring stations reported for Thailand was suspicious (586 reference-grade monitors and 344 low-cost monitors).

The World Air Quality Index project (https://aqicn.org/) is an organization that claims to be non-profit that also compiles and archives air quality data from all around the globe. Its use is discouraged because it is not always clear how environmental agencies' data is accessed. On many occasions, this organization legally obtains data from public applications, but on others, it appears to illegally infiltrate agency servers to obtain the data. The data may be uncertain for the same reasons as with OpenAQ, but with the additional issue that this organization sometimes beats the agencies themselves to publishing the data. All air quality monitoring networks publish their hourly records with a few minutes lag at the end of each hour after the data has been prevalidated and quality assured. This organization is able to get the raw data prior to this procedure in some way. It is common to find web portals of environmental agencies warnings about the use of air quality data and indices posted on *aqicn.org* and other similar online sources.

5.4 Low-Cost Air Quality Monitor Networks

Low-cost air quality monitors have garnered quite a bit of attention in recent years. As the name implies, they are considerably less expensive than regulatory-grade monitors (a few hundreds of dollars or even less versus thousands of dollars) and appear to be much easier to install and operate. This provides potential for building spatially dense networks of monitors (i.e., hyperlocal networks) in order to gain a better understanding of the distribution and temporal variations of air pollutants at scales that are directly relevant to people. This sounds too good to be true, but the reality is that the technology has not yet matured sufficiently to replace the reference air quality monitors currently in use. Compared to reference monitors, low-cost monitors are less sensitive, precise, and chemically specific to the pollutant species of interest. In addition to the fact that their proper operation demands intensive maintenance, which includes frequent checks of their operational status, dynamic corrections due to environmental factors, and periodic indirect calibrations against regulatory-grade monitors. Not to mention their limited useful life; in polluted environments, monitors for measuring gases must be replaced every 6–12 months and particles at least once a year. Users should also consider the costs of sampling capability (e.g., pump), power system (e.g., batteries), housing and weatherproofing, data management (including data collection, post-processing, and storage), and hardware and software for data analysis and visualization (graphs, tables, reports). Although the cost of monitoring networks comprised of low-cost monitors is not as low as one might think, they are less expensive than reference networks, but they may not provide reliable data. Both the US-EPA and WHO have reviewed and evaluated the performance, limitations, and potential applications of low-cost monitors (Duvall et al. 2021a,b; Peltier et al. 2021).

Even though low-cost monitors are not yet sufficiently robust and lack innate quality assurance tools, they can be used under specific conditions to complement existing monitoring systems and develop new applications to better inform on the state of air quality, but only if a robust calibration and validation scheme is implemented to reduce uncertainties in their measurements (e.g., Frederickson et al. 2022; Heimann et al. 2015). This duty must be carried out by environmental authorities and research organizations (see Fig. 5.4). Private companies can provide this service, but they must be overseen by institutions that certify their work. For this reason, it is not recommended to entirely rely on air quality data posted by the public on global networks such as IQAir (https://www.iqair.com/), Purple Air (https://www2.purpleair.com/), and Clarity (https://www.clarity.io/), nor on regional networks such as Nafas in Indonesia (https://nafas.co.id/), SensorForAll in Thailand (https://sensorforall.com/), and PAM Air in Vietnam (https://pamair.org/) other than to get a rough indication of air quality conditions. Monitors purchased and operated by well-meaning citizens or non-governmental organizations cannot guarantee data quality. As non-specialists, they might not necessarily have experience with measurement sciences or air quality monitoring. Nonetheless, the use of these monitors is beneficial for educational purposes and serves as a means of encouraging public engagement

Fig. 5.4 Evaluation of nine different low-cost monitors for measuring $PM_{2.5}$ in Mexico City. The response of low-cost monitors must be tested against a regulatory-grade monitor before they are deployed in the field. This allows calibration curves to be built for each individual monitor under local environmental conditions. Similarly, an intercomparison between two or more monitors enables users to assess the repeatability of their readings and, consequently, gain confidence in their use

and awareness in the search for collective solutions. Readers are invited to review the recommendations from an expert panel convened by the World Meteorological Organization (WMO) on the best practices for using these monitors to improve understanding and management of air quality (WMO 2024).

The study conducted by Singapore' National Environment Agency to assess the potential of a low-cost monitor network deployed across a residential estate to map air pollution at high spatial resolution in a localized area provides a good example of how these sensors can be incorporated into current monitoring systems. Twenty-eight sensors were placed at selected lampposts and on various floors of a building to investigate the dispersion of pollutants within the estate at different heights (NEA 2021). For initial calibration, all sensors were collocated first with a reference monitoring station. The $PM_{2.5}$ monitors worked fine, but the NO_2 sensors did not. The $PM_{2.5}$ data was then used to investigate the impact of thermal turbulence on the vertical distribution of pollutants, demonstrating the potentiality of low-cost monitors to reveal insightful trends and phenomena related to air pollution. Similar to this example, there is a range of peer-reviewed literature describing studies in which low-cost monitors were successfully implemented to answer specific questions in different science domains that rely on information about atmospheric composition (see references in Peltier et al. 2021). For the particular case of Southeast Asia, Lung et al. (2022) reviewed how these sensors have been used in environmental and health research, as well as the challenges they face in being incorporated into air quality management.

References

Acid Deposition Monitoring Network in East Asia (EANET). Data Report 2021. Network Center for EANET. December 2022. https://www.eanet.asia/

Akimoto, H., Sato, K., Sase, H., Dong, Y., Hu, M., Duan, L., Sunwoo, Y., Suzuki, K., Tang, X.: Development of science and policy related to acid deposition in East Asia over 30 years. Ambio 51(8), 1800–1818 (2022). https://doi.org/10.1007/s13280-022-01702-6

Cambaliza, M.O.L., NUS AQ Lab., Latif, M.T., Lestari, P.: Regional and urban air quality in Southeast Asia: Maritime Continent. In: Akimoto, H., Tanimoto, H. (eds.) Handbook of Air Quality and Climate Change. Springer, Singapore (2023). https://doi.org/10.1007/978-981-15-2527-8_68-1

Cambodia Ministry of Environment: Clean Air Plan Cambodia. Kingdom of Cambodia, Phenom Penh, Cambodia. https://www.ccacoalition.org/en/resources/clean-air-plan-cambodia (2021)

Daerah Khusus Ibukota (DKI) and Vital Strategies: Toward Clean Air Jakarta. Environmental Agency of Jakarta, Jakarta, Indonesia. https://www.vitalstrategies.org/ (2019)

Department of Environment and Natural Resources: National Air Quality Status Report 2016–2018. Environmental Management Bureau. Philippine Government. Quezon City, Philippines. https://air.emb.gov.ph/e-library/ (2020)

Department of Environment, Parks, and Recreation: Air quality management in Brunei Darussalam. Ministry of Development. Government of Brunei Darussalam. http://www.env.gov.bn/SitePages/Air%20Quality%20Management%20in%20Brunei%20Darussalam.aspx (2021)

Dhammapala, R.: Analysis of fine particle pollution data measured at 29 US diplomatic posts worldwide. Atmos. Environ. 213, 367–376 (2019). https://doi.org/10.1016/j.atmosenv.2019.05.070

Duvall, R.M., Clements, A.L., Hagler, G., Kamal, A., Kilaru, V., Goodman, L., Frederick, S., Barkjohn, K.K., VonWald, I.,Greene, D., Dye T.: Performance testing protocols, metrics, and target values ozone air sensors. EPA/600/R-20/279. United States Environmental Protection Agency, NC, USA. https://cfpub.epa.gov/si/si_public_record_Report.cfm?dirEntryId=350784&Lab=CEMM (2021a)

Duvall, R.M., Clements, A.L., Hagler, G., Kamal, A., Kilaru, V., Goodman, L., Frederick, S., Barkjohn, K.K., VonWald, I.,Greene, D., Dye T.: Performance testing protocols, metrics, and target values for fine particulate matter air sensors. EPA/600/R-20/280. United States Environmental Protection Agency, NC, USA. https://cfpub.epa.gov/si/si_public_record_Report.cfm?dirEntryId=350785&Lab=CEMM (2021b)

Frederickson, L.B., Sidaraviciute, R., Schmidt, J.A., Hertel, O., Johnson, M.S.: Are dense networks of low-cost nodes better at monitoring air pollution? A case study in Staffordshire. Atmos. Chem. Phys. 22(21), 13949–13965 (2022). https://doi.org/10.5194/acp-22-13949-2022

Giles, D.M., Sinyuk, A., Sorokin, M.G., Schafer, J.S., Smirnov, A., Slutsker, I., Eck, T.F., Holben, B.N., Lewis, J.R., Campbell, J.R., Welton, E.J., Korkin, S.V., Lyapustin, A.I.: Advancements in the Aerosol Robotic Network (AERONET) Version 3 database—automated near-real-time quality control algorithm with improved cloud screening for Sun photometer aerosol optical depth (AOD) measurements. Atmos. Measure. Tech. 12, 169–209 (2019). https://doi.org/10.5194/amt-12-169-2019

Heimann, I., Bright, V.B., McLeod, M.W., Mead, M.I., Popoola, O.A.M., Stewart, G.B., Jones, R.L.: Source attribution of air pollution by spatial scale separation using high spatial density networks of low cost air quality sensors. Atmos. Environ. 113, 10–19 (2015). https://doi.org/10.1016/j.atmosenv.2015.04.057

Holben, B.N., Eck, T.F., Slutsker, I., Tanré, D., Buis, J.P., Setzer, A., Vermote, E., Reagan, J.A., Kaufman, Y., Nakajima, T., Lavenu, F., Jankowiak, I., Smirnov, A.: AERONET—a federated instrument network and data archive for aerosol characterization. Remote Sens. Environ. **66**(1), 1–16 (1998). https://doi.org/10.1016/S0034-4257(98)00031-5

Lao PDR Ministry of Natural Resources and the Environment: Lao Environment Outlook 2012. Lao People's Democratic Republic, Vientiane. https://www.unep.org/resources/report/lao-env ironment-outlook-2012 (2012)

Lao PDR Ministry of Natural Resources and the Environment: National Pollution Control Strategy and Action Plan 2018–2025, with Vision 2030. ADB Greater Mekong Subregion Core Environment Program. Lao People's Democratic Republic, Vientiane. https://www.gms-eoc.org/upl oads/resources/922/attachment/Laos-Pollution-Strategy-Plan-2018-2025-draft.pdf (2017)

Latif, M.T., Dominick, D., Ahamad, F., Khan, M.F., Juneng, L., Hamzah, F.M., Nadzir, M.S.M.: Long term assessment of air quality from a background station on the Malaysian Peninsula. Sci. Total Environ. **482–483**, 336–348 (2014). https://doi.org/10.1016/j.scitotenv.2014.02.132

Lung, S.C.C., Thi Hien, T., Cambaliza, M.O.L., Hlaing, O.M.T., Oanh, N.T.K., Latif, M.T., Lestari, P., Salam, A., Lee, S.Y., Wang, W.C.V., Tsou, M.C.M., Cong-Thran, T., Cruz, M.T., Tantrakarnapa, K.,Othman, M., Roy, S., Dang, T.N., Agustian, D.: Research priorities of applying low-cost $PM_{2.5}$ sensors in Southeast Asian countries. Int. J. Environ. Res. Public Health **19**, 1522 (2022). https://doi.org/10.3390/ijerph19031522

Malaysia's Department of Environment: 2021 Environmental Quality Report. Malaysia Ministry of Environment and Water. https://www.doe.gov.my/en/environmental-quality-report/ (2022)

Martin, R.V., Brauer, M., van Donkelaar, A., Shaddick, G., Narain, U., Dey, S.: No one knows which city has the highest concentration of fine particulate matter. Atmos. Environ. X **3**, 100040 (2019). https://doi.org/10.1016/j.aeaoa.2019.100040

Ministry of Environment and Forestry: Environmental Quality Index 2019. ISBN 978-602-8358-94-1. Government of Indonesia, Jakarta, Indonesia. https://www.menlhk.go.id/site/post/2 (2020)

Molina, L.T., Velasco, E., Retama, A., Zavala, M.: Experience from integrated air quality management in the Mexico City Metropolitan Area and Singapore. Atmosphere **10**(9), 512 (2019). https://doi.org/10.3390/atmos10090512

Moya, M., Madronich, S., Retama, A., Weber, R., Baumann, K., Nenes, A., Castillejos, M., De León, C.P.: Identification of chemistry-dependent artifacts on gravimetric PM fine readings at the T1 site during the MILAGRO field campaign. Atmos. Environ. **45**(1), 244–252 (2011). https://doi. org/10.1016/j.atmosenv.2010.08.059

Myint, Z.L.: Air quality monitoring stations to be established across Myanmar. Myanmar Digital News. https://www.myanmardigitalnewspaper.com/en/air-quality-monitoring-stations-be-established-across-myanmar (2021)

National Environmental Agency (NEA): State of the environment: air and water quality. NEA, Government of Singapore, Singapore. https://www.nea.gov.sg/corporate-functions/resources/ publications (2021)

National Environmental Agency: Environmental Protection Division Annual Report 2018. Government of Singapore. https://www.nea.gov.sg/corporate-functions/resources/publications (2019)

National News Bureau of Thailand: Thailand to tighten air quality safety standards. Royal Thai Government. 15 July. https://thainews.prd.go.th/en/news/detail/TCATG2207151247 33629 (2022)

Oanh, N.T.K., Hlaing, O.M.T., Hien, T.T.: Regional and urban air quality in Mainland Southeast Asia countries. In: Akimoto, H., Tanimoto, H. (eds.) Handbook of Air Quality and Climate Change. Springer, Singapore (2023). https://doi.org/10.1007/978-981-15-2527-8_69-1

Peltier, R.E., Castell, N., Clements, A.L., Hüglin, C., Kroll, J.H., Lung, S.C.C. Ning, Z., Parsons, M., Penza, M., Reisen, F., von Schneidemessee, E.: An update on low-cost sensors for the measurement of atmospheric composition. WMO-No.125, ISBN: 978-92-6-11212-6. World Meteorological Organization, Genova, Switzerland. https://library.wmo.int/index.php?lvl=not ice_display&id=21508#.Y0VGSUxBxhE (2021)

Prime Minister's Office: Agreement on the National Environmental Standards. Water Resources and Environment Administration. No. 2734/PMO.WREA. Vientiane Province, Lao People's Democratic Republic. https://web.mwa.co.th/download/prd01/iDW_standard/LAO_PDR_2009.pdf (2009)

Schultz, M.G., Schröder, S., Lyapina, O., Cooper, O.R., Galbally, I., Petropavlovskikh, I., Von Schneidemesser, E., Tanimoto, H., Elshorbany, Y., Naja, M., Seguel, R.J.: Tropospheric ozone assessment report: database and metrics data of global surface ozone observations. Elementa Sci. Anthropocene 5, 58 (2017). https://doi.org/10.1525/elementa.244

Snider, G., Weagle, C.L., Martin, R.V., Van Donkelaar, A., Conrad, K., Cunningham, D., Gordon, C., Zwicker, M., Akoshile, C., Artaxo, P., et al.: SPARTAN: a global network to evaluate and enhance satellite-based estimates of ground-level particulate matter for global health applications. Atmos. Measure. Tech. 8, 505–521 (2015). https://doi.org/10.5194/amt-8-505-2015

Snider, G., Weagle, C.L., Murdymootoo, K.K., Ring, A., Ritchie, Y., Stone, E., Walsh, A., Akoshile, C., Anh, N.X., Balasubramanian, R., Brook, J., et al.: Variation in global chemical composition of $PM_{2.5}$: emerging results from SPARTAN. Atmos. Chem. Phys. 16(15), 9629–9653 (2016). https://doi.org/10.5194/acp-16-9629-2016

Szykman, J., Swap, R., Lefer, B., Valin, L., Lee, S.C., Fioletov, V., Zhao, X., Davies, J., Williams, D., Abuhassan, N., Shalaby.: Pandora: connecting in-situ and satellite monitoring in support of the Canada–US Air Quality Agreement. In: EM: Air and Waste Management Association's Magazine for Environmental Managers, 2470–4741. https://www.awma.org/empastissues (2019)

United Nations Environment Programme (UNEP). Actions on air quality: policies and programmes for improving air quality around the world. UNEP, Nairobi, Kenia. https://www.unep.org/resources/report/global-report-summary-air-quality-policies-around-world (2021a)

United Nations Environment Programme (UNEP). Major new project to enhance early warning systems for increased climate resilience in Timor-Leste. United Nations Environment Programme. UNEP, Nairobi, Kenia, 8 October. https://www.unep.org/news-and-stories/press-release/major-new-project-enhance-early-warning-systems-increased-climate (2021b)

US Environmental Protection Agency (US-EPA). Monitoring regulations—general monitoring requirements. Ambient monitoring technology information center. US Environmental Protection Agency, Research Triangle Park. https://www.epa.gov/amtic/monitoring-regulations#General (2016)

Vietnam Environment Administration: Enhanced warnings and forecasts to improve air quality. Ministry of Natural Resources and Environment. Government of Vietnam. 27 November. https://vea.gov.vn/sites/en/pages/detail.aspx?$id=442 (2021)

World Bank: Lao PDR Environment Monitor. Washington DC, USA. https://openknowledge.worldbank.org/handle/10986/33929 (2006)

World Bank: Myanmar Country Environmental Analysis. Washington DC, USA. https://documents.worldbank.org/en/publication/documents-reports/documentdetail/464661560176989512/synthesis-report (2019)

World Health Organization (WHO): WHO ambient air quality database 2022. WHO, Geneva, Switzerland. https://www.who.int/data/gho/data/themes/air-pollution/who-air-quality-database (2022b)

World Health Organization (WHO): WHO global air quality guidelines. Particulate matter ($PM_{2.5}$ and PM_{10}), ozone, nitrogen dioxide, sulfur dioxide and carbon monoxide. WHO, Geneva, Switzerland, ISBN 978-92-4-003422-8. https://www.who.int/publications/i/item/9789240034228 (2022a)

World Meteorological Organization (WMO): Integrating low-cost sensor systems and networks to enhance air quality application. Global Atmosphere Watch Programme (GAW) report 293. Geneva, Switzerland. https://library.wmo.int/records/item/68924-integrating-low-cost-sensor-systems-and-networks-to-enhance-air-quality-applications (2024)

Chapter 6
Emission Inventories

Abstract It is critical to know what pollutant species are emitted, where their emission sources are located, and how they evolve over time to design control measures to improve air quality. Emission inventories must provide reliable information to determine the origin of both pollutants directly emitted into the atmosphere and precursor species of secondary pollutants like ozone and an important fraction of fine particles. Emissions must be spatially and temporally distributed, as well as disaggregated by chemical species to be used as input data to run atmospheric chemical-transport models aimed at understanding their impact on ambient concentrations, analyzing emission reduction strategies, and forecasting air quality conditions. However, the construction of emission inventories with such characteristics is still in progress all over Southeast Asia. No country has an emissions inventory that can be used for advanced analysis or air quality modeling. Only four countries have compiled national-level lists of annual emission estimates. Local researchers and international consortia dedicated to developing global and regional emission datasets have taken up the task of building gridded emission inventories. These inventories can be used as a source of information on past and present emissions in the region.

Keywords National emissions inventory · Local emissions inventory · Regional emissions inventory

Emission inventories enable emissions mitigation by providing information on what pollutants are emitted, where their emission sources are located, and how they evolve over time. They are necessary to determine the origin of pollutants and precursors gases so that informed action can be taken to reduce their atmospheric load by addressing their emission sources. Emission inventories are especially important for atmospheric chemical-transport models, which use emission time series as key input data. Nonetheless, despite their relevance for air quality management, the development of emission inventories in Southeast Asia is hampered by a number of obstacles.

Less than half of the region's countries have emission inventories compiled by government agencies. These inventories are constructed at the national or provincial level using bottom-up approaches, i.e., emissions are calculated using detailed statistical analyses of activity data, energy consumption, or any other proxy variable associated with emissions, along with pollutant-specific emission factors. The methodology, level of detail, sectoral coverage, and consistency across time and space differ. Furthermore, none of them provide extensive documentation for assumptions and methods, nor do they include uncertainty estimates.

No country has an official gridded emissions inventory that can be used to run air quality models. There are, however, local academic efforts to build gridded emission inventories at the country or city scale for this purpose, but not at the regional scale across Southeast Asia. Notwithstanding, global and regional emission datasets developed by international consortia such as the Emission Database for Global Atmospheric Research (EDGAR; Crippa et al. 2018, 2020, 2024), which was already introduced in Sect. 3.3, the Community Emissions Data Set (CEDS; Hoesly et al. 2018), and the Regional Emission Inventory in Asia (REAS; Kurokawa and Ohora 2020) can fill in the gaps. These emission datasets are the primary source of emissions data for global air quality research and assessments, such as the State of Global Air Report (HEI 2024).

The following sections of this chapter provide an overview of all available emission inventories for Southeast Asia, including those resulting from international initiatives to develop global and regional inventories, national inventories prepared by government agencies, and inventories compiled for academic purposes by local researchers. Table 6.1 summarizes the details of each emissions inventory, the pollutant species and emission source sectors analyzed, as well as the study area and the spatial scale, are all included. Figure 6.1 compares country-scale annual emissions reported by national emission inventories, whenever they exist, and the three gridded emission datasets used for international air quality assessments.

6.1 Global and Regional Emission Inventories

Global and regional emission datasets, such as EDGAR, REAS, and CEDS, fill the gap of detailed bottom-up emission inventories in most developing countries around the globe. The emission inventories presented below, developed for research and analysis purposes as part of international initiatives, can be used as a source of information on past and present emissions in Southeast Asia.

The Emission Database for Global Atmospheric Research—EDGAR is an independent global database of anthropogenic emissions of air pollutants and greenhouse gases, both present and past (Crippa et al. 2018, 2020, 2024). Emissions are calculated using emission factors that are consistently applied to sets of activity data from all countries. EDGAR provides both emissions as national totals and gridded maps with a resolution of $0.1° \times 0.1°$ for the entire globe, with monthly and annual data. As

Table 6.1 Selected peer-reviewed literature on air pollutant emission estimates at different scales for Southeast Asian countries. All emission inventories have been compiled using a bottom-up approach (i.e., activity data and emission factors)

Reference	Pollutant species[a]	Emission source sectors	Covered area	Spatial resolution	Reference year
Emission Database for Global Atmospheric Research (EDGAR)					
Crippa et al. (2018, 2020)	SO_2, NO_x, CO, NMVOC, NH_3, PM_{10}, $PM_{2.5}$, BC, OC, Hg, CO_2 CH_4, N_2O, HFC, PFCs, SF_6, NF_3, SO_2F_2, CFCs, HCFCs, CCl_4, CH_3Br, CH_3CCl_2	Industry, power generation, fuel production, services, commercial, households, aviation, shipping, railways, vehicular traffic, agriculture, fertilizer application, livestock, biomass burning, wastewater, waste landfill, and waste incineration	Global	$0.1° \times 0.1°$	1970–2022
Community Emissions Data Set (CEDS)					
Hoesly et al. (2018)	NO_x, SO_2, CO, NMVOC, BC, OC, NH_3, CO_2, N_2O, CH_4	Industry, power generation, services, commercial, households, aviation, shipping, railways, vehicular traffic, agriculture, fertilizer application, livestock, wastewater, waste landfill, and waste incineration	Global	$0.5° \times 0.5°$	1750–2019

(continued)

Table 6.1 (continued)

Reference	Pollutant species[a]	Emission source sectors	Covered area	Spatial resolution	Reference year
Regional Emission Inventory in Asia (REAS)					
Kurokawa and Ohora (2020)	SO_2, NO_x, CO, NMVOC, NH_3, PM_{10}, $PM_{2.5}$, BC, OC, CO_2	Industry, power generation, services, commercial, households, railways, vehicular traffic, agriculture, fertilizer application, livestock, wastewater, and latrines	East, Southeast, and South Asia	$0.25° \times 0.25°$	1950–2015
Mosaic Asian Inventory (MIX)					
Li et al. (2024)	SO_2, NO_x, CO, NMVOC, NH_3, PM_{10}, $PM_{2.5}$, BC, OC, CO_2	Industry, power generation, households, transportation, fertilizer application, livestock, wastewater, biomass burning, and shipping	Asia	$0.1° \times 0.1°$	2010–2017
Evaluating the Climate and Air Quality Impacts of Short-Lived Pollutants (ECLIPSE)					
Stohl et al. (2015), Klimont et al. (2017)	NO_x, SO_2, CO, NMVOC, NH_3, PM_1, PM_{10}, $PM_{2.5}$, BC, OC, CO_2, CH_4	Industry, power generation, services, commercial, households, aviation, shipping, railways, vehicular traffic, agriculture, agriculture open burning, livestock, waste landfill, waste incineration, and brick manufacturing	Global	$0.5° \times 0.5°$	1990–2010
Regional emissions from crop residue open burning					
Oanh et al. (2018a, b)	NO_x, SO_2, CO, NMVOC, $PM_{2.5}$, PM_{10}, BC, OC, NH_3, CO_2, N_2O, CH_4, PAHs, BaP, dioxins, OCPs, total chlorine	Agriculture biomass burning	Southeast Asia	$0.1° \times 0.1°$	2010–2015

Table 6.1 (continued)

Reference	Pollutant species[a]	Emission source sectors	Covered area	Spatial resolution	Reference year
Regional carbon emissions from biomass burning					
Shi et al. (2014)	Carbon	Agriculture biomass burning and wildfires	Southeast Asia	5 km × 5 km	2001–2010
Brunei Darussalam					
Dotse et al. (2016)	CO_2, CH_4, N_2O, NO_X, NMVOC, CO, SO_2, PM_{10}	Industry, power generation, households, transportation, agriculture, and waste management	Entire country	0.1° × 0.1°	2001–2012
Indonesia					
Lestari et al. (2022)	NO_X, NMVOC, CO, SO_2, BC, $PM_{2.5}$, PM_{10}	Industry, power generation, services, commercial, households, and vehicular traffic	Jakarta area	2 km × 2 km	2015
Oanh et al. (2018a, b)	NO_x, SO_2, CO, NMVOC, C_4H_6, C_2H_4O, HCHO, C_6H_6, PM, NH_3, CO_2, N_2O, CH_4	Vehicular traffic	Bandung	Entire city	2015
Permadi et al. (2017)	NO_x, SO_2, CO, NMVOC, $PM_{2.5}$, PM_{10}, BC, OC, NH_3, CO_2, N_2O, CH_4	Industry, power generation, services, commercial, households, vehicular traffic, biomass burning, agriculture, and waste	Entire country	Provincial/ district	2007 and 2010
Malaysia					
Azhari et al. (2021)	PM_{10}, $PM_{2.5}$, SO_2, NO_2, NO_x	Vehicular traffic and industry	Kuala Lumpur	1 km × 1 km	2015
Shafie and Mahmud (2020)	PM_{10}, CO, NO_x	Vehicular traffic	Kuala Lumpur	Entire city	2010–2014

(continued)

Table 6.1 (continued)

Reference	Pollutant species[a]	Emission source sectors	Covered area	Spatial resolution	Reference year
Myanmar					
Huy et al. (2020)	NO_x, SO_2, CO, NMVOC, C_4H_6, C_2H_4O, HCHO, C_6H_6, PM, NH_3, CO_2, N_2O, CH_4	Vehicular traffic	Yangon	Entire city	2015
Thailand					
Hongthong et al. (2022)	PM_{10}, $PM_{2.5}$	Biomass burning	Northern Thailand	1 km × 1 km	2012–2016
Junpen et al. (2018)	NO_x, SO_2, CO, PM_{10}, $PM_{2.5}$, BC, OC, CO_2, CH_4	Rice residue open burning	Entire country	12 km × 12 km	2010–2017
Cheewaphongphan et al. (2017)	NO_x, SO_2, NMVOC, CO, PM_{10}, $PM_{2.5}$, BC, OC, NH_3, CO_2, N_2O, CH_4	Vehicular traffic	Bangkok Metropolitan Region	1 km × 1 km	2007–2015
Kanabkaew and Oanh (2011)	NO_x, SO_2, NMVOC, CO, PM_{10}, $PM_{2.5}$, BC, OC, NH_3, CO_2, CH_4	Crop residue open burning	Entire country	0.1° × 0.1°	2007
Vongmahadlek et al. (2009)	NO_x, SO_2, NMVOC, CO, PM_{10}, BC, OC, NH_3	Industry, power generation, vehicular traffic, aviation, navigation, households, incinerators, biomass burning, agriculture, livestock, and biogenic sources	Entire country	1 km × 1 km	2005

(continued)

Table 6.1 (continued)

Reference	Pollutant species[a]	Emission source sectors	Covered area	Spatial resolution	Reference year
Pham et al. (2008)	NO_x, SO_2, NMVOC, CO, BC, OC, NH_3	Industry and power generation	Entire country	Subdistricts	2000 and 2004
Vietnam					
Roy et al. (2022)	NO_x, SO_2, NMVOC, CO, PM_{10}, $PM_{2.5}$, BC, OC, CO_2, N_2O, CH_4	Power generation	Entire country	Regions	2019
Nguyen et al. (2022)	$PM_{2.5}$, BC, OC, NO_x, SO_2, NMVOC, NH_3, CH_4, CO	Agriculture, households, transport, commercial activities, industry, and other sources (solvent use and gas stations)	Hanoi	1 km × 1 km	2017–2018 with projections to 2025 and 2030
Nguyen et al. (2021)	NO_x, SO_2, NMVOC, CO, PM_{10}, $PM_{2.5}$, BC, OC, NH_3, CO_2, N_2O, CH_4	Industry, construction, households, and transportation,	Ho Chi Minh City	1 km × 1 km	2009–2016
Huy et al. (2021)	NO_x, SO_2, NMVOC, CO, PM_{10}, $PM_{2.5}$, BC, OC, NH_3, CO_2, N_2O, CH_4, PAHs	Residential combustion	Red River Delta (Hanoi, Hai Phong, and 8 surrounding provinces)	District	2010–2015
Ho et al. (2019)	NO_x, SO_2, NMVOC, CO, TSP, CH_4, $PM_{2.5}$	Industry, services, construction, biomass burning, households, transportation, airport, railway, and biogenic sources	Ho Chi Minh City	0.5 km × 0.5 km	2017

(continued)

Table 6.1 (continued)

Reference	Pollutant species[a]	Emission source sectors	Covered area	Spatial resolution	Reference year
Lasko and Vadrevu (2018)	PM$_{2.5}$	Rice residue burning	Vietnam and Hanoi's provinces	By region and 4 × 4 km for Hanoi province	2017
Bang et al. (2018)[b]	NO$_x$, SO$_2$, NMVOC, CO, TSP, CH$_4$	Industry, services, construction, biomass burning, households, transportation, airport, and piers	Can Tho City, Mekong Delta	70 km × 70 km	2015

[a] BC, OC, PAHs stand for black carbon, organic carbon, and polyaromatic hydrocarbons, respectively. See the Acronyms section for more abbreviations
[b] It was constructed using both a bottom-up approach and a top-down approach

an example, Fig. 6.2 shows the gridded emissions of four pollutant species for Southeast Asia. To spatially allocate emissions, a geographical database with the location of energy and manufacturing facilities, road networks, shipping routes, human and animal population density, and agricultural land is used as spatial proxy datasets. Emissions are calculated for O$_3$ precursor gases, acidifying gases, primary particles, mercury, direct greenhouse gases, and stratospheric O$_3$ depleting substances.

The Community Emissions Data Set—CEDS provides emissions of anthropogenic reactive gases, aerosols, and greenhouse gases from 1750 to the present, covering the entire globe (Hoesly et al. 2018). Annual emissions are calculated using emission factors and activity data at the country level by fuel and sector, scaled to match country-level inventories where available, and extended over time using a consistent methodology. It excludes agriculture waste burning, which is covered in companion work (van Marle et al. 2017). In a similar manner to EDGAR, emissions are aggregated and processed to provide gridded data with monthly seasonality at $0.5° \times 0.5°$ resolution using spatial proxy data.

The Regional Emission Inventory in Asia—REAS is a third source of emissions data for Southeast Asia (Kurokawa and Ohara 2020). REAS focuses on the long-term historical trend of air pollutants in Asia. Its coverage area encompasses East, Southeast, and South Asia, and its estimates include monthly emissions of major air and climate pollutants with a resolution of $0.25° \times 0.25°$. The majority of the emissions are calculated using emission factors and activity data, but it also makes use of a few previous inventories for specific economic sectors in Japan and Korea. The grid allocation is done following EDGAR's methodology. REAS does not include emissions from aviation and shipping, but it does include NH$_3$ emissions from latrines, which are an important emission source in rural areas of Southeast Asia that has been overlooked by other inventories.

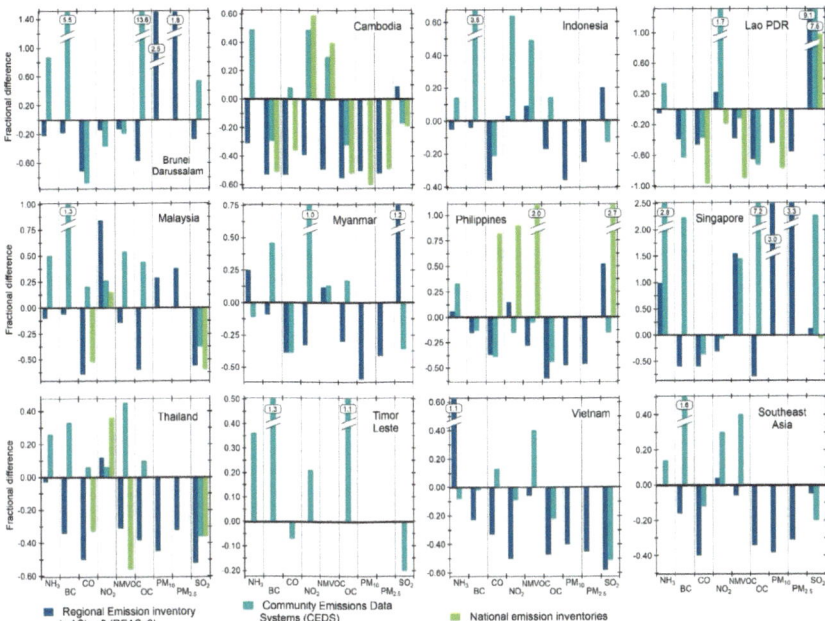

Fig. 6.1 Comparisons of emission inventories at the national level. The Regional Emission inventory in Asia v3 (REASv3), the Community Emissions Data System (CEDS), and national emission estimates from official inventories, when available, are compared with the Emissions Database for Global Atmospheric Research v5 (EDGARv5). Emissions from REASv3, CEDS and EDGARv5 are referenced to 2015, while emissions from national inventories correspond to the country's most recent emission estimates. The ordinate axes represent the percentage by which these inventories overestimate or underestimate the emissions of each country relative to EDGARv5 emissions. References of national emission inventories: Cambodia—2015 (Cambodia Ministry of Environment 2021), Lao PDR—2012, including only industrial emissions (GMS Working Group 2014), Malaysia—2021 (Malaysia's Department of Environment 2022), Philippines—2018 (Department of Environment and Natural Resources 2020), Singapore—2020 (NEA 2021), and Thailand—2016 (Ministry of Natural Resources and Environment 2020)

In support of the Model Inter-Comparison Study for Asia (MICS-Asia) research project, REAS has been taken as a component to build a gridded emissions inventory combining the best available state-of-the-art regional and national emission inventories from across Asia using a mosaic approach. The second version of this inventory, Mosaic Asian Inventory (MIXv2), was recently released (Li et al. 2024). MIXv2 takes the anthropogenic emissions estimated in REAS for Southeast Asia and adds biomass burning and shipping emissions from the Global Fire Emission Database (GFED) and EDGAR, respectively.

Emission datasets developed as part of the Evaluating the Climate and Air Quality Impacts of Short-Lived Pollutants (ECLIPSE) project are another source of emissions data for Southeast Asia (Sthol et al. 2015; Klimont et al. 2017). These datasets

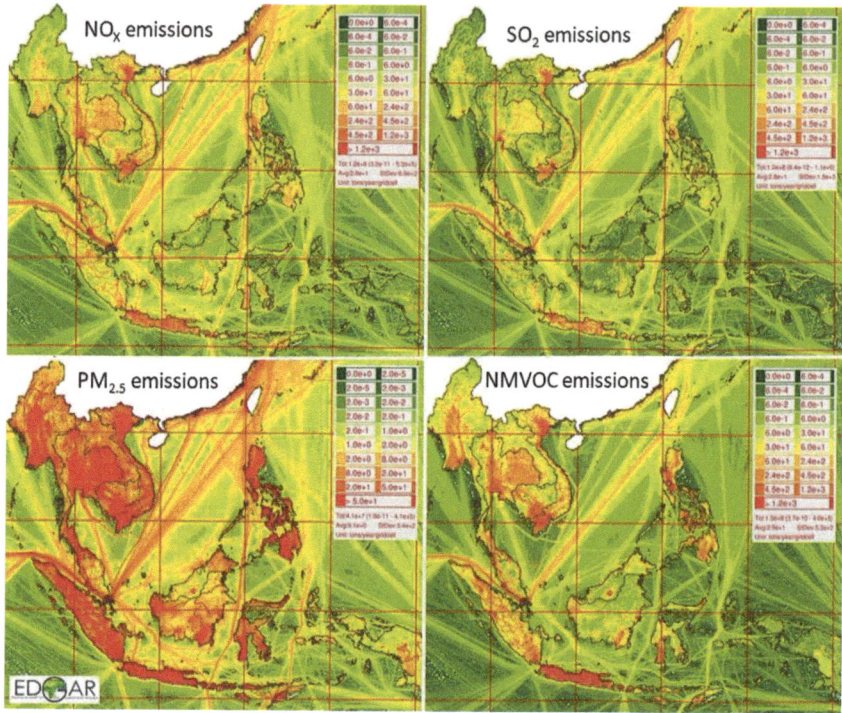

Fig. 6.2 Annual gridded emissions of NO_X, SO_2, $PM_{2.5}$, and nonmethane VOCs (NMVOC) at 0.1° × 0.1° for Southeast Asia in 2018 compiled using EDGAR v6.1 (https://edgar.jrc.ec.europa.eu/)

include air pollutants and greenhouse gas emissions for each country based on emission factors and activity data from international energy and industrial statistics, emission inventories, and data supplied by countries themselves. The emissions were calculated at 0.5° × 0.5° resolution using the Greenhouse gas—Air pollution INteractions and Synergies (GAINS) model (Amann et al. 2011). This model is a tool for comparing climate change mitigation policies and exploring cost-effective control strategies for local and transboundary air pollution.

The uncertainties in the emissions calculated by these four inventories for Southeast Asia may be relatively high when compared to emissions calculated for other parts of the world or Asia itself. This responds to the use of generic emission factors for many sectors due to a lack of locally developed factors, as well as a lack of detailed data for various economic activities, including information on emission controls and technological changes. For example, REASv3 estimates uncertainties ranging from 38% to 232%. Particle emissions, which include PM_{10}, $PM_{2.5}$, black carbon, and organic carbon (OC), have the highest uncertainty (> 100%).

As expected, there are differences between each inventory's estimated emissions. The magnitudes of these differences depend on the pollutant species, but they usually respond to the validity and updating of the emission factors, the degree of detail and representativeness of the activity data used, and the methodologies used to aggregate emissions at various scales. Figure 6.1 shows differences in emissions from EDGARv5, CEVS, and REASv3 for most pollutants in each country ranging from − 50% to 75%. On occasions, the differences exceed 100%, especially for black carbon and SO_2 emissions. For the majority of Southeast Asian countries, black carbon emissions are strongly linked to biomass and fossil fuel combustion in small industries and households, where emission factors and activity data are inaccurate. The differences in the case of SO_2 are most likely due to the emission controls and technology that each inventory takes into account. Power plants and large industries are major SO_2 emission sources, and because they are relevant to each country, uncertainties in activity data are small. It is important to note that the largest differences in inventories occur in Singapore and Brunei Darussalam, the two countries with the highest per capita income in the region (see Table 2.1). This is true for other pollutant species besides black carbon and SO_2.

Except for MIXv2, which includes biomass burning emissions, none of these emission inventories include biogenic and wildfire emissions. The Model of Emissions of Gases and Aerosol from Nature (MEGAN, https://bai.ess.uci.edu/megan) can be used to calculate biogenic emissions. It is an open-source model that estimates the emission of gases and aerosols into the atmosphere from terrestrial ecosystems (Guenther et al. 2012). The Global Fire Emission Database (GFED, https://www.globalfiredata.org/) provides information on emissions from non-agricultural, open biomass burning. It combines satellite data on fire activity and vegetation productivity to calculate gridded fire emissions (van der Werf et al. 2017).

6.2 National Emission Inventories

Cambodia, Malaysia, the Philippines, and Thailand have national-scale emission inventories that report at least three pollutant species, including contributions from industry, power generation, commercial and public services, households, and transportation. The three former countries have emission inventories that are compiled and updated as part of their air quality management activities. These emission inventories are presented in environmental quality reports or air quality programs (Cambodia Ministry of Environment 2021; Malaysia's Department of Environment 2022; Department of Environment and Natural Resources 2020). Thailand's emissions inventory is compiled in preparation for the periodic National Communications to the United Nations Framework Convention on Climate Change (Ministry of Natural Resources and Environment 2020). Lao PDR has an emissions inventory that only accounts for industrial emissions at the province scale (GMS Working Group 2014), whereas Singapore only reports SO_2 emissions from industry and transportation (NEA 2021).

The Philippine's inventory stands out for making provincial-scale emissions publicly available through a user-friendly interactive website (https://air.emb.gov.ph/emission-inventory-2018/). Cambodia's effort to build an emissions inventory that includes pollutant species and contributing sectors that are also included in global and regional inventories is also noteworthy.

All of these emission inventories are compiled on an annual basis using bottom-up approaches and taking into account typical environmental conditions. The veracity of the activity data and the representativeness of the emission factors determine the accuracy of the emission estimates. Activity data is obtained from a variety of sources, most notably demographic and economic statistics compiled by government agencies. While emission factors are taken from international literature and guidelines, in some instances they are adjusted to account for local conditions (e.g., European Environment Agency 2019; US-EPA 2017).

None of these emission inventories appear to have evaluated the accuracy of their emission estimates through direct or indirect observations or by running chemical-transport models. To accurately determine the pollutants' origin, it is necessary to know the uncertainty in the estimated emissions. Nevertheless, when comparing the emissions from these inventories to those estimated by EDGARv5, differences of similar magnitude to those found when comparing the emissions from REASv3 and CEDS at the country level are observed (see Fig. 6.1). This is particularly true for Cambodia, Malaysia, and Thailand. The exception are the emissions reported by the Philippines inventory, which are 2–3 times higher than the emissions reported by the other inventories.

6.3 Local Emission Inventories Compiled for Academic Purposes

Local academia has taken over the task of developing emission inventories for research and air quality modeling to support air quality management in Southeast Asia. Emission inventories of varying scope, coverage, and quality have been built by groups of researchers (see Table 6.1). Many of these inventories have focused on emissions from major sectors such as transportation, power generation, and the burning of agricultural waste at the city, province, or country level. In response to the growing need to run chemical-transport models for the design of better emission control strategies that account for the transport and transformations of pollutants, gridded emission inventories have only recently started to be constructed.

All of these emission inventories have been compiled using bottom-up approaches. Emission factors are taken from literature or derived from models that adjust emission rates to local conditions. For example, the US-EPA-funded International Vehicle Emissions model (IVE, www.issrc.org/ive) has been widely used to determine emission factors for motor vehicles during engine operation and engine startup, taking into account a wide array of fuel types and quality, and vehicle engine technology (e.g., Huy et al. 2020; Oanh et al. 2018a, b; Permadi et al. 2017). Activity data by sector is taken from official statistics as long as they are available. Otherwise, data is taken from statistics compiled by international organizations. In the last instance, specific field campaigns are designed and carried out to collect the required data. For example, for many of the inventories listed in Table 6.1, traffic surveys were conducted to determine vehicle flow, vehicle fleet composition, and travel speeds (e.g., Huy et al. 2020; Ho et al. 2019; Oanh et al. 2018a, b). In the case of agricultural biomass burning emissions, satellite images have been used to quantify both cultivated and burned areas (e.g., Lasko and Vandrevu 2018; Shi et al. 2014).

With the exception of Bang et al. (2018), who evaluated the accuracy of their emission estimates for the Mekong Delta region by applying a numerical model to predict O_3 formation, top-down approaches have not been used to validate bottom-up emission estimates. Top-down approaches quantify emissions based on observations or inverse modeling exercises. Similarly, with the exception of Nguyen et al. (2021), who ran Monte Carlo simulations to analyze uncertainty propagation in emission estimates for Ho Chi Minh City, no other author has formally assessed the uncertainty of estimated emissions. In the best case, only uncertainties based on the author's own judgment, reported by other authors, or suggested by international guidelines have been cited (e.g., Nguyen et al. 2022). Indeed, quantifying uncertainties is difficult because all of the components and assumptions used to build emission inventories vary with time, space, and pollutant species. Consequentially, it is necessary to design studies to evaluate the accuracy of emission inventories through measurements of pollutants chemical speciation, ratios between concentrations of pollutant species, eddy covariance flux measurements, inverse modeling, receptor models, or application of chemical-transport models, among other available techniques (Molina et al. 2002).

References

Amann, M., Bertok, I., Borken-Kleefeld, J., Cofala, J., Heyes, C., Höglund-Isaksson, L., Klimont, Z., Nguyen, B., Posch, M., Rafaj, P., Sander, R., Schöpp, W., Wagner, F., Winiwarter, W.: Cost-effective control of air quality and greenhouse gases in Europe: modeling and policy applications. Environ Model Softw. **26**, 1489–1501 (2011). https://doi.org/10.1016/j.envsoft.2011.07.012

Azhari, A., Halim, N.D.A., Othman, M., Latif, M.T., Juneng, L., Sofwan, N.M., Stocker, J., Johnson, K.: Highly spatially resolved emission inventory of selected air pollutants in Kuala Lumpur's urban environment. Atmos. Pollut. Res. **12**(2), 12–22 (2021). https://doi.org/10.1016/j.apr.2020.10.004

Bang, H.Q., Khue, V.H.N., Tam, N.T., Lasko, K.: Air pollution emission inventory and air quality modeling for Can Tho City, Mekong Delta, Vietnam. Air Qual. Atmos. Health **11**, 35–47 (2018). https://doi.org/10.1007/s11869-017-0512-x

Cambodia Ministry of Environment: Clean Air Plan Cambodia. Kingdom of Cambodia, Phenom Penh, Cambodia. https://www.ccacoalition.org/en/resources/clean-air-plan-cambodia (2021)

Cheewaphongphan, P., Junpen, A., Garivait, S., Chatani, S.: Emission inventory of on-road transport in Bangkok metropolitan region (BMR) development during 2007 to 2015 using the GAINS model. Atmosphere **8**(9), 167 (2017). https://doi.org/10.3390/atmos8090167

Crippa, M., Guizzardi, D., Muntean, M., Schaaf, E., Dentener, F., van Aardenne, J. A., Monni, S., Doering, U., Olivier, J. G. J., Pagliari, V., Janssens-Maenhout, G.: Gridded emissions of air pollutants for the period 1970–2012 within EDGAR v4.3.2. Earth Syst. Sci. Data **10**, 1987–2013 (2018). https://doi.org/10.5194/essd-10-1987-2018

Crippa, M., Guizzardi, D., Pagani, F., Schiavina, M., Melchiorri, M., Pisoni, E., Graziosi, F., Muntean, M., Maes, J., Dijkstra, L., Van Damme, M., Clarisse, L., and Coheur, P.: Insights on the spatial distribution of global, national and sub-national GHG emissions in EDGARv8.0, Earth Syst. Sci. Data **16**(6), 2811–2830, 2024 https://doi.org/10.5194/essd-16-2811-2024

Crippa, M., Solazzo, E., Huang, G., Guizzardi, D., Koffi, E., Muntean, M., Schieberle, C., Friedrich, R., Janssens-Maenhout, G.: High resolution temporal profiles in the Emissions Database for Global Atmospheric Research. Sci Data **7**, 121 (2020). https://doi.org/10.1038/s41597-020-0462-2

Department of Environment and Natural Resources: Emissions Inventory 2018. Environmental Management Bureau. Quezon City, Philippines. https://air.emb.gov.ph/emission-inventory-2018/ (2020)

Dotse, S.Q., Dagar, L., Petra, M.I., De Silva, L.C.: Evaluation of national emissions inventories of anthropogenic air pollutants for Brunei Darussalam. Atmos. Environ. **133**, 81–92 (2016). https://doi.org/10.1016/j.atmosenv.2016.03.024

European Environment Agency: EMEP/EEA air pollutant emission inventory guidebook 2019. https://doi.org/10.2800/293657. European Environment Agency, European Union, Copenhagen, Denmark. https://www.eea.europa.eu/ (2019)

Greater Mekong Subregion (GMS) Working Group: Estimating industrial pollution in Lao PDR. Final report. GMS Core Environment Program. GMS Environment Operation Center, Bangkok, Thailand. https://www.gms-eoc.org/ (2014)

Guenther, A.B., Jiang, X., Heald, C.L., Sakulyanontvittaya, T., Duhl, T., Emmons, L.K., Wang, X.: The model of emissions of gases and aerosols from nature version 2.1 (MEGAN2.1): an extended and updated framework for modeling biogenic emissions. Geoscientific Model Develop. **5**, 1471–1492 (2012). https://doi.org/10.5194/gmd-5-1471-2012

Health Effects Institute (HEI).: State of Global Air 2024. Special Report. Health Effects Institute, Boston, MA (2024). https://www.stateofglobalair.org/

Ho, Q.B., Vu, H.N.K., Nguyen, T.T., Nguyen, T.T.H., Nguyen, T.T.T.: A combination of bottom-up and top-down approaches for calculating of air emission for developing countries: a case of Ho Chi Minh City, Vietnam. Air Qual. Atmos. Health **12**, 1059–1072 (2019). https://doi.org/10.1007/s11869-019-00722-8

Hoesly, R.M., Smith, S.J., Feng, L., Klimont, Z., Janssens-Maenhout, G., Pitkanen, T., Seibert, J.J., Vu, L., Andres, R.J., Bolt, R.M., Bond, T.C., Dawidowski, L., Kholod, N., Kurokawa, J.I., Li, M., Liu, L., Lu, Z., Moura, M.C.P., O'Rourke, P.R., Zhang, Q.: Historical (1750–2014) anthropogenic emissions of reactive gases and aerosols from the Community Emissions Data System (CEDS). Geoscientific Model Develop. **11**, 369–408 (2018). https://doi.org/10.5194/gmd-11-369-2018

Hongthong, A., Nanthapong, K., Kanabkaew, T.: Biomass burning emission inventory of multi-year PM_{10} and $PM_{2.5}$ with high temporal and spatial resolution of Northern Thailand. SicenceAsia **48**, 302–309 (2022). https://doi.org/10.2306/scienceasia1513-1874.2022.040

Huy, L.N., Oanh, N.T.K., Htut, T.T., Hlaing, O.M.T.: Emission inventory for on-road traffic fleets in Greater Yangon, Myanmar. Atmos. Pollut. Res. **11**(4), 702–713 (2020). https://doi.org/10.1016/j.apr.2019.12.021

Huy, L.N., Oanh, N.T.K., Phuc, N.H., Nhung, C.P.: Survey-based inventory for atmospheric emissions from residential combustion in Vietnam. Environ. Sci. Pollut. Res. **28**, 10678–10695 (2021). https://doi.org/10.1007/s11356-020-11067-6

Junpen, A., Pansuk, J., Kamnoet, O., Cheewaphongphan, P., Garivait, S.: Emission of air pollutants from rice residue open burning in Thailand, 2018. Atmosphere **9**(11), 449 (2018). https://doi.org/10.3390/atmos9110449

Kanabkaew, T., Kim Oanh, N.T.: Development of spatial and temporal emission inventory for crop residue field burning. Environ. Model. Assess. **16**, 453–464 (2011). https://doi.org/10.1007/s10666-010-9244-0

Klimont, Z., Kupiainen, K., Heyes, C., Purohit, P., Cofala, J., Rafaj, P., Borken-Kleefeld, J., Schöpp, W.: Global anthropogenic emissions of particulate matter including black carbon. Atmos. Chem. Phys. **17**(14), 8681–8723 (2017). https://doi.org/10.5194/acp-17-8681-2017

Kurokawa, J., Ohara, T.: Long-term historical trends in air pollutant emissions in Asia: Regional Emission inventory in Asia (REAS) version 3. Atmos. Chem. Phys. **20**(21), 12761–12793 (2020). https://doi.org/10.5194/acp-20-12761-2020

Lasko, K., Vadrevu, K.: Improved rice residue burning emissions estimates: accounting for practice-specific emission factors in air pollution assessments of Vietnam. Environ. Pollut. **236**, 795–806 (2018). https://doi.org/10.1016/j.envpol.2018.01.098

Lestari, P., Arrohman, M.K., Damayanti, S., Klimont, Z.: Emissions and spatial distribution of air pollutants from anthropogenic sources in Jakarta. Atmos. Pollut. Res. **13**(9), 101521 (2022). https://doi.org/10.1016/j.apr.2022.101521

Li, M., Kurokawa, J., Zhang, Q., Woo, J.-H., Morikawa, T., Chatani, S., Lu, Z., Song, Y., Geng, G., Hu, H., Kim, J., Cooper, O.R., McDonald, B.C.: MIXv2: a long-term mosaic emission inventory for Asia (2010–2017). Atmos. Chem. Phys. **24**(7), 3925–3952 (2024). https://doi.org/10.5194/acp-24-3925-2024

Malaysia's Department of Environment: 2021 Environmental Quality Report. Malaysia Ministry of Environment and Water. https://www.doe.gov.my/en/environmental-quality-report/ (2022)

Ministry of Natural Resources and Environment: Thailand third biennial update report. Office of Natural Resources and Environmental Policy and Planning. Bangkok, Kingdom of Thailand. https://unfccc.int/documents/267629 (2020)

Molina, J.M., Molina, L.T., West, J., Sosa, G., Sheinbaum, C., San Martini, F., Zavala, M.A., McRae, G.: Air pollution science in the MCMA: Understandings source-receptor relationships through emissions inventories, measurements, and modeling. In: Molina, L.T., Molina, M.J. (eds.) Air Quality in the Mexico Megacit, pp. 137–212. Kluwer Academic Publishers, Dordrecht, The Netherlands, ISBN 978-1-4020-0507-7 (2002. https://doi.org/10.1007/978-94-010-0454-1#toc

National Environmental Agency (NEA): Air Pollution. NEA, Government of Singapore, Singapore. https://www.nea.gov.sg/our-services/pollution-control/air-pollution/air-quality (2021)

Nguyen, T.H., Hung, N.T., Nagashima, T., Lam, Y.F., Doan, Q.V., Kurokawa, J., Chatani, S., Derdouri, A., Cheewaphongphan, P., Khan, A., Niyogi, D.: Development of current and future high-resolution gridded emission inventory of anthropogenic air pollutants for urban air quality studies in Hanoi, Vietnam. Urban Climate **46**, 101334 (2022). https://doi.org/10.1016/j.uclim.2022.101334

Nguyen, T.T.Q., Takeuchi, W., Misra, P., Hayashida, S.: Emission mapping of key sectors in Ho Chi Minh City, Vietnam, using satellite-derived urban land use data. Atmos. Chem. Phys. **21**(4), 2795–2818 (2021). https://doi.org/10.5194/acp-21-2795-2021

Oanh, N.T.K., Huy, L.N., Permadi, D.A., Zusman, E., Nakano, R., Nugroho, S.B., Lestari, P., Sofyan, A.: Assessment of urban passenger fleet emissions to quantify climate and air quality co-benefits resulting from potential interventions. Carbon Manage. **9**, 367–381 (2018a). https://doi.org/10.1080/17583004.2018.1500790

Oanh, N.T.K., Permadi, D.A., Hopke, P.K., Smith, K.R., Dong, N.P., Dang, A.N.: Annual emissions of air toxics emitted from crop residue open burning in Southeast Asia over the period of 2010–2015. Atmos. Environ. **187**, 163–173 (2018b)

Permadi, D.A., Sofyan, A., Oanh, N.T.K.: Assessment of emissions of greenhouse gases and air pollutants in Indonesia and impacts of national policy for elimination of kerosene use in cooking. Atmos. Environ. **154**, 82–94 (2017). https://doi.org/10.1016/j.atmosenv.2017.01.041

Pham, T.B.T., Manomaiphiboon, K., Vongmahadlek, C.: Development of an inventory and temporal allocation profiles of emissions from power plants and industrial facilities in Thailand. Sci. Total. Environ. **397**, 103–118 (2008). https://doi.org/10.1016/j.scitotenv.2008.01.066

Roy, S., Lam, Y.F., Chan, J.C., Hung, N.T., Fu, J.S.: Evaluation of Vietnam air emissions and the impacts of revised power development plan (PDP7 rev) on spatial changes in the thermal power sector. Atmos. Pollut. Res. **13**(7), 101454 (2022). https://doi.org/10.1016/j.apr.2022.101454

Shafie, S.H.M., Mahmud, M.: Urban air pollutant from motor vehicle emissions in Kuala Lumpur, Malaysia. Aerosol Air Qual. Res. **20**(12), 2793–2804 (2020). https://doi.org/10.4209/aaqr.2020. 02.0074

Shi, Y., Sasai, T., Yamaguchi, Y.: Spatio-temporal evaluation of carbon emissions from biomass burning in Southeast Asia during the period 2001–2010. Ecol. Model. **272**, 98–115 (2014). https://doi.org/10.1016/j.ecolmodel.2013.09.021

Stohl, A., Aamaas, B., Amann, M., Baker, L.H., Bellouin, N., Berntsen, T.K., Boucher, O., Cherian, R., Collins, W., Daskalakis, N., Dusinska, M., Eckhardt, S., Fuglestvedt, J.S., Harju, M., Heyes, C., Hodnebrog, Ø., Hao, J., Im, U., Kanakidou, M., Klimont, Z., Kupiainen, K., Law, K.S., Lund, M.T., Maas, R., MacIntosh, C.R., Myhre, G., Myriokefalitakis, S., Olivié, D., Quaas, J., Quennehen, B., Raut, J.-C., Rumbold, S.T., Samset, B.H., Schulz, M., Seland, Ø., Shine, K.P., Skeie, R.B., Wang, S., Yttri, K.E., Zhu, T.: Evaluating the climate and air quality impacts of short-lived pollutants. Atmos. Chem. Phys. **15**, 10529–10566 (2015). https://doi.org/10.5194/acp-15-10529-2015

United States—Environmental Protection Agency (US-EPA): Emissions inventory guidance for implementation of ozone and particulate matter national ambient air quality standards (NAAQS) and regional haze regulations. EPA-454/B-17-002. US-EPA, NC US. https://www.epa.gov/air-emissions-inventories (2017)

van der Werf, G.R., Randerson, J.T., Giglio, L., Van Leeuwen, T.T., Chen, Y., Rogers, B.M., Mu, M., van Marle, M.J., Morton, D.C., Collatz, G.J., Yokelson, R.J., Kasibhatla, P.S.: Global fire emissions estimates during 1997–2016. Earth Syst. Sci. Data **9**, 697–720 (2017). https://doi.org/10.5194/essd-9-697-2017

van Marle, M.J.E., Kloster, S., Magi, B.I., Marlon, J.R., Daniau, A.-L., Field, R.D., Arneth, A., Forrest, M., Hantson, S., Kehrwald, N.M., Knorr, W., Lasslop, G., Li, F., Mangeon, S., Yue, C., Kaiser, J.W., van der Werf, G.R.: Historic global biomass burning emissions for CMIP6 (BB4CMIP) based on merging satellite observations with proxies and fire models (1750–2015). Geoscientific Model Develop. **10**, 3329–3357 (2017). https://doi.org/10.5194/gmd-10-3329-2017

Vongmahadlek, C., Bich Thao, P.T., Satayopas, B., Thongboonchoo, N.: A compilation and development of spatial and temporal profiles of high-resolution emissions inventory over Thailand. J. Air Waste Manage. Assoc. **59**, 845–856 (2009). https://doi.org/10.3155/1047-3289.59.7.845

Chapter 7
Air Quality Modeling and Forecasting

Abstract Air quality models help explain how pollutants are transported, transformed, and dispersed in the atmosphere. They are useful for analyzing past air pollution episodes and future scenarios that account for changes in emissions and climate, as well as assessing the effectiveness of emission mitigation strategies. A well-tuned air quality model can be run on a daily basis to forecast air quality conditions in the following days, informing the public of potentially harmful levels of atmospheric pollution. However, despite the valuable information that air quality models can provide, no Southeast Asian country has implemented a modeling system as part of its air quality management. Air quality models for Southeast Asia have been run by local and international researchers. This chapter describes the Eulerian, Lagrangian, and hybrid models, as well as machine-learning algorithms, that they have used to answer specific questions about the impact of specific emission sources, such as biomass burning, on individual subregions or the entire region, or to investigate the atmospheric processes that drive air pollution at the province or city scale. It also reviews global and regional air quality forecast models developed and operated by international consortia as alternatives to air quality forecasting in Southeast Asia.

Keywords Atmospheric chemical transport model · Air quality forecasting · Eulerian model · Lagrangian dispersion model · Machine learning

Numerical simulations can be used to represent how air pollutants are transported, transformed, and dispersed in the atmosphere, as well as how they affect air quality. Atmospheric chemical transport models (air quality models in short) are useful for quantifying the impact of anthropogenic and natural emissions on the ambient concentrations and deposition (wet and dry) of atmospheric pollutants under prevailing meteorological conditions. Air quality models can also be used to analyze past air pollution events and future scenarios taking into account changes in emissions and climate, as well as to assess the effectiveness of emission reduction strategies. For these reasons, air quality modeling is a fundamental component of air quality management, becoming indispensable for designing and evaluating emission control-based public policies. Air quality models are also required for forecasting purposes. Many environmental agencies now run air quality models as part of their daily management operations to warn citizens about potential air pollution episodes and to implement control measures before they occur. Several articles have examined the application, progress, and importance of air quality modeling (e.g., Baklanov and Zhang 2020; Mathur et al. 2017; Solazzo et al. 2012; Kukkonen et al. 2012).

This chapter reviews the global and regional air quality forecast models developed and operated by international consortia that cover Southeast Asia, as well as academic studies that have run air quality models over the region entirely or partially. The details of both global and regional forecasting models and models used for academic purposes are provided in Table 7.1.

7.1 Air Quality Forecast Models

Despite the importance of operational air quality models, no Southeast Asian country has integrated a modeling system into its air quality management. This is partially due to the lack of spatially and temporally distributed high-resolution emission inventories and the limited air quality and meteorological data from monitoring stations required to validate the model results.

Frequent smoke-haze episodes triggered by biomass burning from aggressive deforestation and agricultural practices throughout the region repeatedly indicate the need for advanced monitoring and forecasting systems that include air quality monitoring stations at ground level and high-frequency satellite imagery, as well as air quality models at the local and regional scales (Velasco and Rastan 2015). Given the small number of monitoring stations and the limitations of satellites to make observations in the typically cloudy skies of Southeast Asia, the use of numerical models to simulate the transport, dispersion, and chemical transformation of the pollutants embedded in the smoke-haze over the region has been proposed on multiple occasions as an alternative approach to forecast the arrival of these noxious plumes. However, a proper modeling system for forecasting air quality has not yet been implemented. As part of the ASEAN Agreement on Transboundary Haze Pollution signed in 2002 (ASEAN 2002), Singapore runs the open software HYSPLIT Atmospheric Transport and Dispersion Modeling System developed by the National Oceanic and Atmospheric Administration (NOAA) https://www.arl.noaa.gov/hysplit/; Stein et al. 2015) on a daily basis to forecast the occurrence of transboundary haze from wildfires in the region. The HYSPLIT model simulates the haze dispersion from fire hotspots detected by NOAA's satellites in Southeast Asia (https://asmc.asean.org/). Efforts have been made to tailor a more robust Lagrangian dispersion modeling system for real-time prediction based on high-resolution numerical weather simulations and satellite-based active-fire detection (Hansen et al. 2019; Hertwig et al. 2015). Nevertheless, its operational implementation is unclear.

The forecast services at the global and regional scales provided by the Copernicus Atmospheric Monitoring Service (CAMS) of the European Union's Earth Observation Program (https://atmosphere.copernicus.eu/), the Goddard Earth

Table 7.1 Selected peer-reviewed literature on air quality modeling in Southeast Asia

Copernicus Atmospheric Monitoring Service (CAMS)[a]

References	Model	Model type	Pollutant species	Sectors	Covered area	Spatial resolution	Modeled period
Peuch et al. (2022)	Integrated Forecasting System (IFS cycle 47R1) of the European Centre for Medium-Range Weather Forecasts (ECMWF). It includes a numerical weather prediction system coupled with numerical and assimilation systems to forecast aerosols (IFS-ARS, Rémy et al. 2022) and reactive trace gases (IFS-CB05, Flemming et al. 2015)	Integrated forecasting system including satellite and ground-based observations together with meteorological numerical models and atmospheric chemical composition through data assimilation and reanalysis	Available output maps of: O_3, NO_2, CO, SO_2, $PM_{2.5}$ HCHO, and AOD Modeled chemical species: 56 reactive trace gases and 7 types of aerosols[f]	Anthropogenic, natural and biogenic sources, biomass burning, sea-salt, wind-blown desert dust, and lightening NO	Global	40 km × 40 km	2003 onwards

(continued)

Table 7.1 (continued)

References	Model	Model type	Pollutant species	Sectors	Covered area	Spatial resolution	Modeled period
NASA Goddard Earth Observing System Composition Forecast (GEOS-CF) system[b]							
Keller et al. (2021)	Hybrid model between an online weather and chemistry assimilation system (GEOS Earth System Model, GEOS-ESM, the GEOS Data Assimilation System, GEOS-DAS) and an offline Chemical Transport Model application (GEOS-Chem, Long et al. 2015), with a development pathway toward a fully coupled forecasting system with integrated trace gases and aerosols	Integrated forecasting system including satellite and ground-based observations together with meteorological numerical models and atmospheric chemical composition through data assimilation and reanalysis	Available output maps of: O_3, NO_2, CO, SO_2, and $PM_{2.5}$ Modeled chemical species: 250 reactive trace gases and 7 types of aerosols[f]	Anthropogenic, natural and biogenic sources, biomass burning, sea-salt, wind-blown desert dust, and lightening NO (Keller et al. 2014)	Global	25 km × 25 km	2018 onwards

(continued)

Table 7.1 (continued)

References	Model	Model type	Pollutant species	Sectors	Covered area	Spatial resolution	Modeled period
Whole Atmosphere Community Climate Model (WACCM) of the US National Center for Atmospheric Research (NCAR)[c]							
Gettelman et al. (2019)	Modeling system based on offline simulations of the Model for Ozone and Related chemical Tracers (MOZART) Chemistry Mechanism in the Community Earth System Model Version 2 (CESM2) driven by GEOS meteorological forecasts and CAMS anthropogenic emission estimates	Integrated forecasting system including satellite and ground-based observations together with meteorological numerical models and atmospheric chemical composition through data assimilation and reanalysis	Available output maps of: O_3, NO_x, CO, SO_2, $PM_{2.5}$, HCHO, and dust Modeled chemical species: 151 reactive trace gases and 7 types of aerosols[f]	Anthropogenic, natural and biogenic sources, biomass burning, sea-salt, wind-blown desert dust, and lightening NO	Global	100 km × 100 km	2018 onwards
System for Integrated Modeling of Atmospheric Composition (SILAM)[d]							
Sofiev et al. (2015)	Modeling system that incorporates both Eulerian and Lagrangian transport routines, 8 chemico-physical transformation modules (basic acid chemistry and secondary aerosol formation, O_3 formation in the troposphere and the stratosphere, radioactive decay, aerosol dynamics in the air, and pollen transformations), 3- and 4-dimensional variational data assimilation modules	Global-to-meso-scale dispersion model including data assimilation developed for atmospheric composition, air quality, and emergency decision support applications, as well as for inverse dispersion problem solution	Available output maps of: O_3, NO, NO_2, CO, SO_2, PM_{10}, $PM_{2.5}$, dust, and AOD Modeled chemical species: 33 reactive trace gases and 6 types of aerosols[f]	Anthropogenic, natural and biogenic sources, biomass burning, sea-salt, wind-blown desert dust, pollen, and lightening NO	Global	35 km × 35 km	2011 onwards

(continued)

Table 7.1 (continued)

References	Model	Model type	Pollutant species	Sectors	Covered area	Spatial resolution	Modeled period
Mekong Air Quality Explorer (MAQE)[e]							
Gupta et al. (2021)	Hybrid monitoring system based on NASA's GEOS global forecast system, and NASA's Modern-Era Retrospective analysis for Research and Applications v2 (MERRA-2) reanalysis data, and surface $PM_{2.5}$ data from regulatory air quality monitoring stations into a machine-learning algorithm	Data assimilation system based on GEOS forecasted fields coupled with a machine-learning algorithm to generate bias corrected air quality forecasts using aerosols data from satellite and ground-based monitoring stations	$PM_{2.5}$	–	Thailand and Lower Mekong region	$0.5° \times 0.625°$	2020 onwards
Regional air quality models							
Huang et al. (2022)	Weather Research and Forecasting model (WRF) v4.3 and the Community Multiscale Air Quality (CMAQ) v5.3.3 two-way coupled model	Mesoscale forecasting Eulerian model coupled with a chemical model	$PM_{2.5}$ and O_3	Biomass burning	Mainland Southeast Asia and Southern China	$1° \times 1°$	Mar–Apr 2015

(continued)

Table 7.1 (continued)

References	Model	Model type	Pollutant species	Sectors	Covered area	Spatial resolution	Modeled period
Wang et al. (2022)	GEOS-Chem global 3-D chemical transport model v12.1.1 driven by assimilated meteorology from NASA's MERRA-2	Atmospheric chemistry model driven by meteorological data	O_3	Anthropogenic and biogenic sources, and biomass burning	Maritime and mainland Southeast Asia	$2° \times 2.5°$	2005–2016
Reddington et al. (2021)	WRF v3.7.1 coupled with a chemistry component modeling system (WRF-Chem) and the Global Model of Aerosol Processes (GLOMAP)	Mesoscale forecasting Eulerian model coupled with a chemical model	$PM_{2.5}$	Anthropogenic and biogenic sources, and biomass burning	Mainland Southeast Asia and Southeastern China	30 km × 30 km	2003–2015
Annuaylojaroen et al. (2020)	HYSPLIT model integrated into WRF v3.8.1	Lagrangian dispersion model	$PM_{2.5}$	Biomass burning	Mainland Southeast Asia, part of East Asia, and India	$0.01° \times 0.01°$	Mar–Apr 2012
Vongruang and Pimonsree (2020)	WRF v3.8.1 and CMAQ v4.7.1 modeling system	Mesoscale forecasting Eulerian model coupled with a chemical model	PM_{10}	Biomass burning	Mainland Southeast Asia	27 km × 27 km	Mar 2012
Nguyen et al. (2019)	WRF v3.4 and CMAQ v5.0.2 modeling system	Mesoscale forecasting Eulerian model coupled with a chemical model	NO_X, NMVOCS, $PM_{2.5}$, and O_3	Anthropogenic and biogenic sources, and biomass burning	Mainland Southeast Asia	24 km × 24 km	2006–2015 Projections: 2015–2046

(continued)

Table 7.1 (continued)

References	Model	Model type	Pollutant species	Sectors	Covered area	Spatial resolution	Modeled period
Hansen et al. (2019)	Numerical Atmospheric-dispersion Modeling Environment (NAME) III v6.5 with chemistry scheme off integrated into the UK Met Office Unified Model (MetUM)	Lagrangian dispersion model	PM_{10}	Biomass burning	Maritime and mainland Southeast Asia	17 km × 17 km	2010–2015
Kiely et al. (2019)	WRF v3.7 and WRF-Chem modeling system	Mesoscale forecasting Eulerian model coupled with a chemical model	PM_{10}, $PM_{2.5}$, and AOD	Anthropogenic and biogenic sources, and biomass burning	Borneo, Sumatra and Peninsular Malaysia	30 km × 30 km	Aug–Oct 2015
Lee et al. (2019)	WRF v3.6.1 and WRF-Chem modeling system	Mesoscale forecasting Eulerian model coupled with a chemical model	PM_{10}, $PM_{2.5}$, O_3, and CO	Anthropogenic sources including shipping focusing on the use of coal	Maritime and mainland Southeast Asia	36 km × 36 km	2006–2008
Lee et al. (2018)	WRF v3.6.1 and WRF-Chem modeling system	Mesoscale forecasting Eulerian model coupled with a chemical model	PM_{10}, $PM_{2.5}$, O_3, and CO	Anthropogenic sources and biogenic sources, and biomass burning	Maritime and mainland Southeast Asia	36 km × 36 km	2002–2008

(continued)

Table 7.1 (continued)

References	Model	Model type	Pollutant species	Sectors	Covered area	Spatial resolution	Modeled period
Crippa et al. (2016) Health-impact	WRF v3.5 and CAM-Chem modeling system	Mesoscale forecasting Eulerian model coupled with a chemical model	PM_{10}, and $PM_{2.5}$	Anthropogenic and biogenic sources, and biomass burning	Maritime and part of mainland Southeast Asia	10 km × 10 km	Sep–Dec 2015
Hertwig et al. (2015)	NAME III v6.3 integrated into MetUM	Lagrangian dispersion model	PM_{10}	Biomass burning	Maritime and mainland Southeast Asia	4° × 4°	Mar–Apr 2013, Jun 2013, Mar–Apr 2014
Dong and Fu (2015)	WRF v3.4 and CMAQ v5.0.1 modeling system	Mesoscale forecasting Eulerian model coupled with a chemical model	O_3, NO_2, SO_2, CO, PM_{10}, and $PM_{2.5}$	Anthropogenic and biogenic sources, and biomass burning	Mainland Southeast Asia and East Asia	1° × 1°	Mar–Apr 2005–2010
Reddington et al. (2014)	Reading Offline Trajectory (ROTRAJ) Lagrangian 3D model integrated into GLOMAP	Lagrangian dispersion model	$PM_{2.5}$	Biomass burning	Maritime and mainland Southeast Asia	2.8° × 2.8°	2004–2007
Amnuaylojaroen et al. (2014)	WRF v3.3 and WRF-Chem modeling system	Mesoscale forecasting Eulerian model coupled with a chemical model	O_3, and CO	Anthropogenic and biogenic sources, and biomass burning	Maritime and mainland Southeast Asia	36 km × 36 km	Mar 2008 and Dec 2008

(continued)

Table 7.1 (continued)

References	Model	Model type	Pollutant species	Sectors	Covered area	Spatial resolution	Modeled period
Huang et al. (2013)	WRF and CMAQ modeling system	Mesoscale forecasting Eulerian model coupled with a chemical model	CO, $PM_{2.5}$, aerosol composition (SO_4^{2-}, NO_3^-, NH_4^+, K^+), OC, and EC	Biomass burning	Mainland Southeast Asia and part of East Asia	$1° \times 1°$	Mar–Apr 2006
Fu et al. (2012)	WRF v3.1.1 and CMAQ v4.6 modeling system	Mesoscale forecasting Eulerian model coupled with a chemical model	O_3, NO_2, CO, and $PM_{2.5}$	Anthropogenic and biogenic sources, and biomass burning	Mainland Southeast Asia and East Asia	$1° \times 1°$	Mar 2006
Brunei Darussalam							
Dotse et al. (2018)	A suite of machine-learning algorithms	24 h forecasting	PM_{10}	–	Brunei Darussalam	4 locations	2009–2013
Indonesia							
Langmann and Heil (2004)	3D regional atmospheric chemistry model REgional MOdel (REMO)	Eulerian model including chemistry	Total PM, and PM_{10}	Biomass burning and peat fires	Indonesia	$0.5° \times 0.5°$	1997–1998
Malaysia							
Balogun and Tella (2022)	A suite of machine-learning algorithms	O_3 prediction of past events	O_3	–	West coast area of Peninsular Malaysia	Ten locations	2012–2016
Hyer and Chew (2010)	Navy Aerosol and Analysis System (NAAPS)	Offline aerosol Eulerian model	PM_{10}	Biomass burning	Peninsular Malaysia, Sumatra and Borneo	$1° \times 1°$	Sep–Oct 2006

(continued)

Table 7.1 (continued)

References	Model	Model type	Pollutant species	Sectors	Covered area	Spatial resolution	Modeled period
Thailand							
Thongthammachart et al. (2023)	WRF v3.8 and CMAQ v5.2.1 modeling system coupled with the Light Gradient Boosting Machine (LightGBM) algorithm	Mesoscale forecasting Eulerian model coupled with a machine-learning algorithm	$PM_{2.5}$	Anthropogenic and biogenic sources, and biomass burning	Central Thailand (Bangkok, Nonthaburi, Samutprakan, and Pathumthani)	5 km × 5 km	2019
Annuaylojaroen et al. (2022)	Nested Regional Climate Model with Chemistry (NRCM-Chem)	Mesoscale forecasting Eulerian model coupled with a chemical model	$PM_{2.5}$	Anthropogenic and biogenic sources, and biomass burning under the Representative Concentration Pathway (RCP) 8.5	Northern Thailand	1 km × 1 km	1990–1999, 2020–2029
Khodmanee and Annuaylojaroen (2021)	WRF v3.8.1 and WRF-Chem modeling system	Mesoscale forecasting Eulerian model coupled with a chemical model	O_3, NO_2, and CO	Anthropogenic and biogenic sources, and biomass burning	Northern Thailand	10 km × 10 km	Mar 2014

(continued)

Table 7.1 (continued)

References	Model	Model type	Pollutant species	Sectors	Covered area	Spatial resolution	Modeled period
Vietnam							
Rakholia et al. (2022)	A suite of machine-learning algorithms	Forecasting (1 to 24 h) system based	$PM_{2.5}$	–	Ho Chi Minh City	Six locations	Feb–Dec 2021
Minh et al. (2021)	WRF v4.1.1 with Advanced Research (WRF-ARW) coupled with a suite of machine-learning algorithms	Mesoscale forecasting (24, 48, 72 h) Eulerian model coupled with machine-learning algorithms	$PM_{2.5}$	–	Ho Chi Minh City	1 km × 1 km	Sep 2020, and Jun. 2021
Ho et al. (2020)	Air Pollution Model developed by Commonwealth Scientific and Industrial Research Organization (TAPM-CTM)	Mesoscale forecasting Eulerian model coupled with a chemical model	O_3, NO_2, SO_2, and CO	Anthropogenic and biogenic sources	Ho Chi Minh City	2.5 km × 2.5 km	2017, 2025, 2030
Bang et al. (2018)	Finite Volume Model—Transport and Photochemistry Mesoscale model (FVM-TAPOM)	Finite volume model coupled with a transport and photochemistry mesoscale model	O_3	Anthropogenic sources, and biomass burning	Can Tho City, Mekong Delta	70 km × 70 km	2015

[a] CAMS global atmospheric composition forecast portal: https://atmosphere.copernicus.eu/global-forecast-plots
[b] GEOS-CF portal: https://gmao.gsfc.nasa.gov/weather_prediction/GEOS-CF/
[c] WACCM portal: https://www2.acom.ucar.edu/acresp/forecasts-and-near-real-time-nrt-products
[d] SILAM portal: https://silam.fmi.fi/aqforecast.html
[e] MAQE portal: https://servir.adpc.net/tools/mekong-air-quality-explorer
[f] Types of modeled aerosols: desert dust, sea salt, organic matter, black carbon, sulfate, nitrate, and ammonium aerosol

Observing Composition Forecasting (GEOS-CF) system developed by NASA's Global Modeling and Assimilation Office (GMAO) (https://gmao.gsfc.nasa.gov/gmao_mission/), the Whole Atmosphere Community Climate Model (WACCM) operated by the US National Center for Atmospheric Research (NCAR) (https://www2.acom.ucar.edu/), and the System for Integrated modeling of Atmospheric coMposition (SILAM) developed by the Finnish Meteorological Institute (http://silam.fmi.fi/), are alternatives for air quality forecasting in Southeast Asia.

Copernicus Atmospheric Monitoring Service provides 5-day hourly forecasts of selected atmospheric pollutants, including aerosols and gaseous pollutants of primary and secondary origin, for the entire globe twice per day. Through data assimilation and reanalysis, CAMS combines satellite and ground-based observations with numerical models of the atmosphere to estimate the abundance of trace gases and aerosols in the troposphere and, in the case of O_3, the stratosphere (Peuch et al. 2022). Figure 7.1 shows examples of one-day forecasts for selected pollutants throughout Southeast Asia. The maps show the spatial distribution of each pollutant species and how plumes from, for example, large cities are transported by prevailing winds across the region. Plumes from major cities such as Jakarta, Kuala Lumpur, Bangkok, and Manila can be seen clearly in the forecast. In the case of NO_2, plumes can also be seen from Chiang Mai, Singapore, Hanoi, Ho Chi Minh City, and Surabaya. Similarly, the impact of Java's coal power plants is evident in the case of SO_2. Formaldehyde (HCHO) is a product of volatile organic compounds (VOCs) oxidation in the atmosphere and is thus used as a proxy for VOC emissions. On a regional scale, vegetation and wildfires are the major emission sources of HCHO, while vehicular traffic and industries are also significant sources in cities (Bauwens et al. 2016). In this context and as shown in Fig. 7.1e, tropical forests in Southeast Asia are important sources of HCHO, as is biomass burning, which did not appear to occur on the date used as an example. In terms of aerosols, CAMS provides total AOD forecasts as well as forecasts for dust, sea-salt, biomass burning, and sulfate aerosols individually. Readers can find high-quality visualizations of CAMS forecast products at *windy.com* or *earth.nullschool.net*, and SILAM visualizations at *ventusky.com*. These are private websites that offer visualization services for weather and air quality forecasts on mobile devices and the web.

Goddard Earth Observing Composition Forecasting produces daily 5-day forecasts that include the spatial distribution of key pollutants (O_3, NO_2, CO, SO_2, and $PM_{2.5}$) at the surface and the chemical composition of the troposphere and stratosphere on a global and regional scale (Keller et al. 2021). Similar to CAMS, GEOS-CF uses satellite and ground-based observations in conjunction with numerical models of the atmosphere to generate air quality forecasts. Air pollution forecasts for the Southeast Asia region are under the *Seven Seas* region tag. NCAR produces daily 10-day global forecasts using WACCM, a modeling system based on offline simulations of the Model for Ozone and Related chemical Tracers (MOZART) chemistry mechanism in the Community Earth System Model version 2 (CESM2), which is driven by GEOS meteorological forecasts and CAMS anthropogenic emission estimates (Gettelman et al. 2019). Similarly, the Finnish Meteorological Institute provides

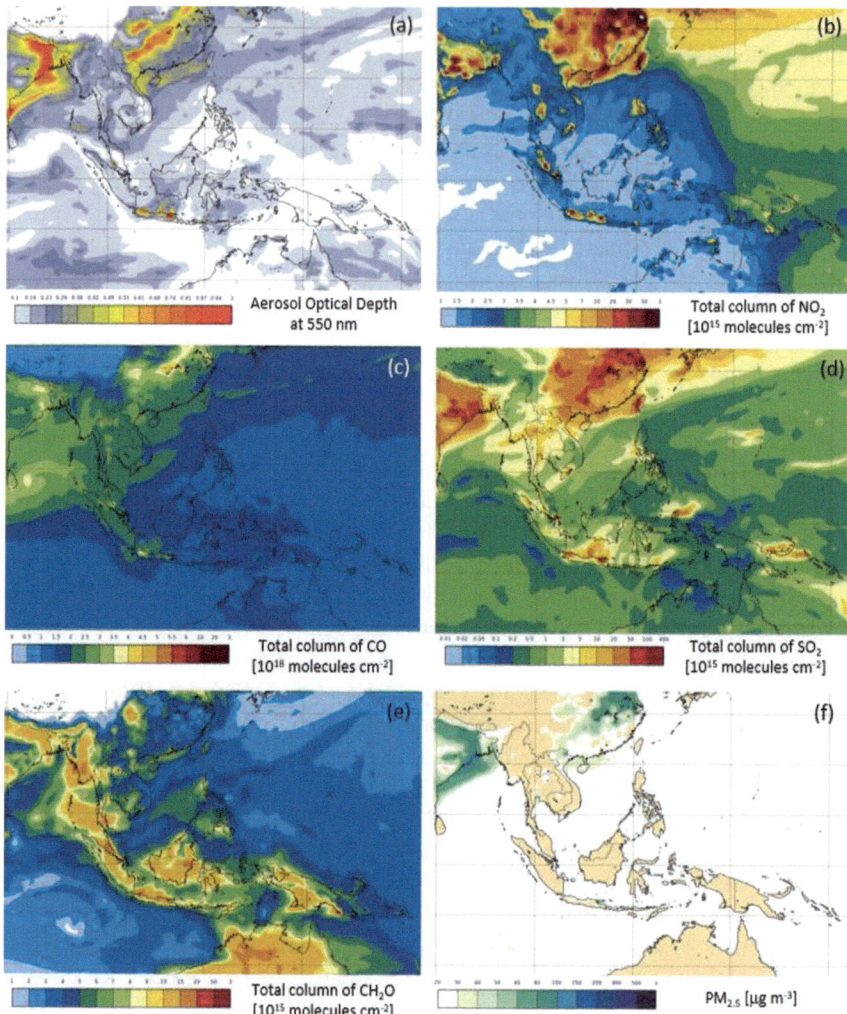

Fig. 7.1 Examples of forecasts of aerosol optical depth (AOD) (**a**), total columns of NO$_2$, CO, SO$_2$, and HCHO (**b–e**), and PM$_{2.5}$ mass concentration (**f**) provided by the Copernicus Atmospheric Monitoring Service (CAMS) for Tuesday 20 December 2022 with a lead time of 24 hours (https:// atmosphere.copernicus.eu/). Except for PM$_{2.5}$ mass concentration, forecast plots for regions other than Europe display air pollutants in terms of molecular densities per unit area along an imaginary column through the entire atmosphere (e.g., 10^{15} molecules of NO$_2$ per cm^{-2}) as opposed to mass or volumetric concentrations at the surface (e.g., μg m^{-3} or ppb of NO$_2$). Nonetheless, users can retrieve data at the surface, 137 model levels, and 25 pressure levels through the CAMS website

daily 4-day global forecasts of atmospheric composition using SILAM, a global-to-meso-scale dispersion model developed for atmospheric composition, air quality, emergency decision support applications, and inverse dispersion problem solution (Sofiev et al. 2015).

Although not exactly an air quality forecast model, it is worth mentioning the Mekong Air Quality Explorer (MAQE, https://aq-tracker-servir.adpc.net/), a hybrid monitoring system leveraging geospatial technology to monitor and forecast air pollution across Thailand and the Lower Mekong region. It is a joint initiative from the Asian Disaster Preparedness Center (ADPC) and the US NASA. MAQE produces 3-day air quality forecasts based on NASA's GEOS global forecast model and reanalysis data of aerosols and meteorology to train a machine-learning-based bias correction algorithm to improve estimates accuracy in Southeast Asia (Gupta et al. 2021).

The air quality forecasts from these modeling systems can differ due to differences in the underlying meteorological fields, observational constraints, data assimilation procedures, chemical and aerosol species, chemical mechanisms, increased model resolution, and pollutants emission data (Huijnen et al. 2019; Xian et al. 2019), without ignoring the uncertainties of the models themselves, which can lead to persistent systematic and random errors (Im et al. 2015a,b). Therefore, their outputs should only be used as an indicator of air quality conditions in coming days. It will be necessary to implement air quality systems that integrate multi-system analysis and downscaling models adapted to local conditions and capable of operating from the regional to the urban scale if they are to be used as an early warning system for potential air pollution episodes. This is acknowledged by the Global Air Quality Forecasting and Information System (GAFIS) project, a new initiative of the World Meteorological Organization's Global Atmospheric Watch (GAW) program that aims to build a platform to support the use of global air quality forecasting systems in a harmonized and standardized manner tailored to the needs of society (https://community.wmo.int/activity-areas/gaw/science-for-services/gafis). To learn about the fundamentals, implementation, operation, and application of air quality forecast systems, it is advised to review the material compiled by the World Meteorological Organization on the subject (WMO 2020).

7.2 Regional and Local Air Quality Models for Academic Purposes

Similarly to the case of emission inventories, academia has taken on the responsibility of developing regional and local air quality models. Researchers from international consortia and local groups have run mesoscale Eulerian models coupled with chemical models to evaluate the impact of anthropogenic emissions and biomass burning throughout the region (see Table 7.1). Most of these modeling exercises have made use of gridded emissions data from the global and regional emission datasets described in the previous chapter (i.e., EDGAR, CEDS, and REA), and biomass burning emissions data from the Global Fire Emissions Database (GFED, https://www.globalfiredata.org/). They have all used the Weather Research and Forecasting model (WRF, https://www.mmm.ucar.edu/models/wrf) to recreate the meteorological fields of past air quality episodes. Some groups have used the chemical module of WRF (WRF-Chem, https://ruc.noaa.gov/wrf/wrf-chem/), while others have used the Community Multiscale Air Quality model (CMAQ, https://www.epa.gov/cmaq) to simulate atmospheric chemical reactions. All of them, to a greater or lesser extent, have used air quality data from regulatory-grade monitoring stations at ground level, aerosol optical depth (AOD) data from AERONET, and satellite observations to validate their output products. Because of their regional focus, the spatial resolution is relatively coarse (grid cells 15–35 km on a side), not fine enough to address air pollution on a local scale (e.g., city). These modeling studies typically use the resulting spatial distributions of $PM_{2.5}$, PM_{10}, CO, and O_3 to assess different air pollution aspects in the region.

Some of these studies have sought to quantify pollutant concentrations in poorly monitored or unmonitored regions of Southeast Asia, as well as to assess the contribution of different emission sources, particularly biomass burning, to the total load of pollutants during specific smoke-haze episodes (Vongruang and Pimonsree 2020; Kiely et al. 2019; Aouizerats et al. 2015; Fu et al. 2012), as well as for longer time frames that include periods affected and unaffected by biomass burning events, with the aim of investigating inter-seasonal and interannual variations, and transport pathways (Lee et al. 2018; Dong and Fu 2015; Huang et al. 2013). The impact on public health and premature deaths caused by exposure to poor air quality caused by forest and peatland fires has also been investigated (Reddington et al. 2021; Crippa et al. 2016). Huang et al. (2022) investigated the impact of air pollution on regional meteorology caused by changes in the optical properties of aerosols resulting from such episodes. These models have also been used to evaluate the impact of emission projections based on different climate change scenarios (Nguyen et al. 2019), as well as the impact of future fuel consumption scenarios (Lee et al. 2019). Finally, the impact of changes in emission trends and the use of different emission datasets in the models has been analyzed trough sensitivity simulations (Wang et al. 2022; Amnuaylojaroen et al. 2014).

Another group of studies has used Lagrangian dispersion models to assess the impact of biomass burning events across the region at specific locations (Amnuaylojaroen et al. 2020; Hertwig et al. 2015), as well as the impact of long term fire emissions from different Southeast Asia's regions (Hansen et al. 2019; Reddington et al. 2014). All of these studies have focused on increases in aerosol load, analyzing changes in PM_{10} and $PM_{2.5}$ concentrations. Fire emissions data have been obtained from GFED or derived from satellite-based active-fire detection, while modeled particle loads have been evaluated using ambient particle concentrations reported by air quality monitoring stations, and AOD data from AERONET. With the exception of Reddington et al. (2014), no other study based on this type of models has included detailed aerosol processes, simulating the evolution of aerosols from emission or production to removal from the atmosphere.

A third group of air quality modeling studies includes studies that cover smaller areas, ranging from the country scale to the city scale (see Table 7.1). Most of these studies have used mesoscale Eulerian models coupled with chemical mechanisms, similar to those described above, but with the option of using other mesoscale and chemical models from the literature as well (e.g., Amnuaylojaroen et al. 2022; Khodmanee and Amnuaylojaroen 2021; Ho et al. 2020; Bang et al. 2018; Hyer and Chew 2010; Langman and Heil 2004). These studies either have used gridded emission datasets from the literature or built their own tailored to the model they use. Similarly, all of these studies have used air quality data from regulatory monitoring stations within the modeling region to calibrate and validate their simulations.

A final group of modeling studies involves the use of machine-learning algorithms to reproduce or forecast ambient concentrations of pollutants under specific meteorological conditions in locations where air quality observations are available (e.g., Belogun and Tella 2022; Rakholia et al. 2022; Minh et al. 2021; Dotse et al. 2018). The application of data-driven models can provide information on the correlation between air pollutants and meteorological parameters, and can thus be used to complement and correct for bias in the outputs of deterministic air quality models (i.e., atmospheric chemical transport models), thereby reducing uncertainty. They can also be used to downscale model outputs to finer spatial resolutions, to fill data gaps in time series, and to develop parameterizations based on the local features (e.g., weather, land use, urban morphology, clusters of emission sources, etc.) of monitoring sites acting as data providers (Liu et al. 2022; Liao et al. 2020). This is how Thongthammachart et al. (2023) used a non-linear machine-learning algorithm integrated into the WRF-CMAQ modeling system to develop a land use regression model to predict daily ambient $PM_{2.5}$ levels in Thailand.

Machine-learning methods have also proven to be effective data mining tools for identifying data patterns and solving problem scenarios. In Southeast Asia, for example, Li et al. (2022) used one of these algorithms to remove the confounding effects of weather conditions on air pollution levels in order to determine the effect of the COVID-19 lockdown in Singapore's air quality, whereas Moosavi et al. (2015) developed a data-driven mathematical model to map particle pollution at street level with fine spatial resolution (100 m × 100 m) across one district of the same city.

Indeed, machine-learning algorithms open up new avenues for air quality research, but they cannot replace the use of deterministic air quality models. Machine-learning algorithms operate as black boxes, and it is difficult or impossible to determine how the environmental predictors produce the results. Furthermore, they are highly sensitive to data quality and quantity, and can produce biased results. It is critical not to abuse them. They can return irrational results if their assumptions and basic requirements (e.g., data distribution, scaling, and standardization) are ignored, as well as if the fundamentals of atmospheric chemistry and physics are ignored in the results' interpretation. Data-driven models can assist decision making, but decisions should not be based only on them. Because the use of these models alone does not result in a complete understanding of the factors that cause air pollution, they should be seen as only as complements to existing air quality management tools.

References

Amnuaylojaroen, T., Barth, M.C., Emmons, L.K., Carmichael, G.R., Kreasuwun, J., Prasitwat-tanaseree, S., Chantara, S.: Effect of different emission inventories on modeled ozone and carbon monoxide in Southeast Asia. Atmos. Chem. Phys. **14**, 12983–13012 (2014). https://doi.org/10.5194/acp-14-12983-2014

Amnuaylojaroen, T., Inkom, J., Janta, R., Surapipith, V.: Long range transport of Southeast Asian $PM_{2.5}$ pollution to Northern Thailand during high biomass burning episodes. Sustainability **12**(23), 10049 (2020). https://doi.org/10.3390/su122310049

Amnuaylojaroen, T., Surapipith, V., Macatangay, R.C.: Projection of the near-future $PM_{2.5}$ in Northern Peninsular Southeast Asia under RCP8.5. Atmosphere **13**(2), 305 (2022). https://doi.org/10.3390/atmos13020305

Aouizerats, B., van der Werf, G.R., Balasubramanian, R., Betha, R.: Importance of transboundary transport of biomass burning emissions to regional air quality in Southeast Asia during a high fire event. Atmos. Chem. Phys. **15**, 363–373 (2015). https://doi.org/10.5194/acp-15-363-2015

Association of Southeast Asian Nations (ASEAN): ASEAN Agreement on Transboundary Haze Pollution. ASEAN Secretariat, Jakarta, Indonesia. http://haze.asean.org (2002)

Baklanov, A., Zhang, Y.: Advances in air quality modeling and forecasting. Global Transitions **2**, 261–270 (2020). https://doi.org/10.1016/j.glt.2020.11.001

Balogun, A.L., Tella, A.: Modelling and investigating the impacts of climatic variables on ozone concentration in Malaysia using correlation analysis with random forest, decision tree regression, linear regression, and support vector regression. Chemosphere **299**, 134250 (2022). https://doi.org/10.1016/j.chemosphere.2022.134250

Bang, H.Q., Khue, V.H.N., Tam, N.T., Lasko, K.: Air pollution emission inventory and air quality modeling for Can Tho City, Mekong Delta, Vietnam. Air Qual. Atmos. Health **11**, 35–47 (2018). https://doi.org/10.1007/s11869-017-0512-x

Bauwens, M., Stavrakou, T., Müller, J.-F., De Smedt, I., Van Roozendael, M., van der Werf, G.R., Wiedinmyer, C., Kaiser, J.W., Sindelarova, K., Guenther, A.: Nine years of global hydrocarbon emissions based on source inversion of OMI formaldehyde observations. Atmos. Chem. Phys. **16**, 10133–10158 (2016). https://doi.org/10.5194/acp-16-10133-2016

Crippa, P., Castruccio, S., Archer-Nicholls, S., Lebron, G.B., Kuwata, M., Thota, A., Sumin, S., Butt, E., Wiedinmyer, C., Spracklen, D.V.: Population exposure to hazardous air quality due to the 2015 fires in Equatorial Asia. Sci. Rep. **6**, 37074 (2016). https://doi.org/10.1038/srep37074

Dong, X., Fu, J.S.: Understanding interannual variations of biomass burning from Peninsular Southeast Asia, part I: model evaluation and analysis of systematic bias. Atmos. Environ. **116**, 293–307 (2015). https://doi.org/10.1016/j.atmosenv.2015.06.026

Dotse, S.Q., Petra, M.I., Dagar, L., De Silva, L.C.: Application of computational intelligence techniques to forecast daily PM_{10} exceedances in Brunei Darussalam. Atmos. Pollut. Res. **9**(2), 358–368 (2018). https://doi.org/10.1016/j.apr.2017.11.004

Flemming, J., Huijnen, V., Arteta, J., Bechtold, P., Beljaars, A., Blechschmidt, A.-M., Diamantakis, M., Engelen, R.J., Gaudel, A., Inness, A., Jones, L., Josse, B., Katragkou, E., Marecal, V., Peuch, V.-H., Richter, A., Schultz, M.G., Stein, O., Tsikerdekis, A.: Tropospheric chemistry in the integrated forecasting system of ECMWF. Geoscientific Model Develop. **8**, 975–1003 (2015). https://doi.org/10.5194/gmd-8-975-2015

Fu, J.S., Hsu, N.C., Gao, Y., Huang, K., Li, C., Lin, N.-H., Tsay, S.-C.: Evaluating the influences of biomass burning during 2006 BASE-ASIA: a regional chemical transport modeling. Atmos. Chem. Phys. **12**, 3837–3855 (2012). https://doi.org/10.5194/acp-12-3837-2012

Gettelman, A., Mills, M.J., Kinnison, D.E., Garcia, R.R., Smith, A.K., Marsh, D.R., Tilmes, S., Vitt, F., Bardeen, C.G., McInerny, J. Liu, H.L., Solomon, S.C., Polvani, L.M., Emmons, L.K., Lamarque, J.F., Richter, J.H., Glanville, A.S., Cacmeister, J.T., Phillips, A.S., Neale, R.B., Simpson, I.R., DuVivier, A.K., Hodzic, A., Randel, W.: The whole atmosphere community climate model version 6 (WACCM6). J. Geophys. Res. Atmosp. **124**(23), 12380–12403 (2019). https://doi.org/10.1029/2019JD030943

Gupta, P., Zhan, S., Mishra, V., Aekakkararungroj, A., Markert, A., Paibong, S., Chishtie, F.: Machine learning algorithm for estimating surface PM$_{2.5}$ in Thailand. Aerosol Air Qual. Res. **21**, 210105 (2021). https://doi.org/10.4209/aaqr.210105

Hansen, A.B., Witham, C.S., Chong, W.M., Kendall, E., Chew, B.N., Gan, C., Hort, M.C., Lee, S.Y.: Haze in Singapore—source attribution of biomass burning PM$_{10}$ from Southeast Asia. Atmos. Chem. Phys. **19**(8), 5363–5385 (2019). https://doi.org/10.5194/acp-19-5363-2019

Hertwig, D., Burgin, L., Gan, C., Hort, M., Jones, A., Shaw, F., Witham, C., Zhang, K.: Development and demonstration of a Lagrangian dispersion modeling system for real-time prediction of smoke haze pollution from biomass burning in Southeast Asia. J. Geophys. Res. Atmos. **120**(24), 12605–12630 (2015). https://doi.org/10.1002/2015JD023422

Ho, B.Q., Vu, K.H.N., Nguyen, T.T., Nguyen, H.T.T., Ho, D.M., Nguyen, H.N., Nguyen, T.T.T.: Study loading capacities of air pollutant emissions for developing countries: a case of Ho Chi Minh City, Vietnam. Sci. Rep. **10**, 5827 (2020). https://doi.org/10.1038/s41598-020-62053-4

Huang, K., Fu, J.S., Hsu, N.C., Gao, Y., Dong, X., Tsay, S.C., Lam, Y.F.: Impact assessment of biomass burning on air quality in Southeast and East Asia during BASE-ASIA. Atmos. Environ. **78**, 291–302 (2013). https://doi.org/10.1016/j.atmosenv.2012.03.048

Huang, Y., Lu, X., Fung, J.C., Wong, D.C., Li, Z., Chen, Y., Chen, W.: Investigating Southeast Asian biomass burning by the WRF-CMAQ two-way coupled model: emission and direct aerosol radiative effects. Atmos. Environ. **294**, 119521 (2022)

Huijnen, V., Pozzer, A., Arteta, J., Brasseur, G., Bouarar, I., Chabrillat, S., Christophe, Y., Doumbia, T., Flemming, J., Guth, J., Josse, B., Karydis, V.A., Marécal, V., Pelletier, S.: Quantifying uncertainties due to chemistry modelling—evaluation of tropospheric composition simulations in the CAMS model (cycle 43R1). Geoscientific Model Develop. **12**, 1725–1752 (2019). https://doi.org/10.5194/gmd-12-1725-2019

Hyer, E.J., Chew, B.N.: Aerosol transport model evaluation of an extreme smoke episode. Atmos. Environ. **44**, 1422–1427 (2010). https://doi.org/10.1016/j.atmosenv.2010.01.043

Im, U., Bianconi, R., Solazzo, E., Kioutsioukis, I., Badia, A., Balzarini, A., Baró, R., Bellasio, R., Brunner, D., Chemel, C., Curci, G., Flemming, J., Forkel, R., Giordano, L., Jiménez-Guerrero, P., Hirtl, M., Hodzic, A., Honzak, L., Jorba, O., Knote, C., Kuenen, J. J., Makar, P. A., Manders-Groot, A., Neal, L., Pérez, J.L., Pirovano, G., Pouliot, G., San Jose, R., Savage, N., Schroder, W., Sokhi, R.S., Syrakov, D., Torian, A., Tuccella, P., Werhahn, J., Wolke, R., Yahya, K., Zabkar, R., Zhang, Y., Zhang, J., Hogrefe, C., Galmarini, S.: Evaluation of operational on-line-coupled regional air quality models over Europe and North America in the context of AQMEII phase 2. Part I: Ozone. Atmos. Environ. **115**, 404–420 (2015a). https://doi.org/10.1016/j.atmosenv.2014.09.042

Im, U., Bianconi, R., Solazzo, E., Kioutsioukis, I., Badia, A., Balzarini, A., Baró, R., Bellasio, R., Brunner, D., Chemel, C., Curci, G., Denier van der Gon, H., Flemming, J., Forkel, R., Giordano, L., Jiménez-Guerrero, P., Hirtl, M., Hodzic, A., Honzak, L., Jorba, O., Knote, C., Makar, P.A., Manders-Groot, A., Neal, L., Pérez, J.L., Pirovano, G., Pouliot, G., San Jose, R., Savage, N., Schroder, W., Sokhi, R.S., Syrakov, D., Torian, A., Tuccella, P., Wang, K., Werhahn, J., Wolke, R., Zabkar, R., Zhang, Y., Zhang, J., Hogrefe, C., Galmarini, S.: Evaluation of operational online-coupled regional air quality models over Europe and North America in the context of AQMEII phase 2. Part II: Particulate matter. Atmos. Environ. **115**, 421–441 (2015b). https://doi.org/10.1016/j.atmosenv.2014.08.072

Keller, C.A., Long, M.S., Yantosca, R.M., Da Silva, A.M., Pawson, S., Jacob, D.J.: HEMCO v1.0: a versatile, ESMF-compliant component for calculating emissions in atmospheric models. Geoscientific Model Develop. **7**, 1409–1417 (2014). https://doi.org/10.5194/gmd-7-1409-2014

Keller, C.A., Knowland, K.E., Duncan, B.N., Liu, J., Anderson, D.C., Das, S., Lucchesi, R.A., Lundgren, E.W., Nicely, J.M., Nielsen, E., Ott, L.E., Saunders, E., Strode, S.A., Wales, P.A., Jacob. D. J., Pawson, S.: Description of the NASA GEOS composition forecast modeling system GEOS-CF v1. 0. J. Adv. Model. Earth Syst. **13**(4), e2020MS002413 (2021). https://doi.org/10.1029/2020MS002413

Khodmanee, S., Amnuaylojaroen, T.: Impact of biomass burning on ozone, carbon monoxide, and nitrogen dioxide in Northern Thailand. Front. Environ. Sci. **9**, 27 (2021). https://doi.org/10. 3389/fenvs.2021.641877

Kiely, L., Spracklen, D.V., Wiedinmyer, C., Conibear, L., Reddington, C.L., Archer-Nicholls, S., Lowe, D., Arnold, S.R., Knote, C., Khan, M.F., Latif, M.T., Kuwata, M., Budisulistiorini, S.H., Syaufina, L.: New estimate of particulate emissions from Indonesian peat fires in 2015. Atmos. Chem. Phys. **19**, 11105–11121 (2019). https://doi.org/10.5194/acp-19-11105-2019

Kukkonen, J., Olsson, T., Schultz, D.M., Baklanov, A., Klein, T., Miranda, A.I., Monteiro, A., Hirtl, M., Tarvainen, V., Boy, M., Peuch, V.H., Poupkou, A., Kioutsioukis, I., Finardi, S., Sofiev, M., Sokhi, R., Lehtinen, K.E.J., Karatzas, K., San José, R., Astitha, M., Kallos, G., Schaap, M., Reimer, E., Jakobs, H., Eben, K.: A review of operational, regional-scale, chemical weather forecasting models in Europe. Atmos. Chem. Phys. **12**, 1–87 (2012). https://doi.org/10.5194/ acp-12-1-2012

Langmann, B., Heil, A.: Release and dispersion of vegetation and peat fire emissions in the atmosphere over Indonesia 1997/1998. Atmos. Chem. Phys. **4**, 2145–2160 (2004). https://doi.org/ 10.5194/acp-4-2145-2004

Lee, H.H., Iraqui, O., Wang, C.: The impact of future fuel consumption on regional air quality in Southeast Asia. Sci. Rep. **9**, 2648 (2019). https://doi.org/10.1038/s41598-019-39131-3

Lee, H.H., Iraqui, O., Gu, Y., Yim, S.H.-L., Chulakadabba, A., Tonks, A.Y.-M., Yang, Z., Wang, C.: Impacts of air pollutants from fire and non-fire emissions on the regional air quality in Southeast Asia. Atmos. Chem. Phys. **18**, 6141–6156 (2018). https://doi.org/10.5194/acp-18-6141-2018

Li, Y., Zhu, Y., Tan, J.Y.K., Teo, H.C., Law, A., Qu, D., Luo, W.: The impact of COVID-19 on NO_2 and $PM_{2.5}$ levels and their associations with human mobility patterns in Singapore. Annals of GIS (2022). https://doi.org/10.1080/19475683.2022.2121855

Liao, Q., Zhu, M., Wu, L., Pan, X., Tang, X., Wang, Z.: Deep learning for air quality forecasts: a review. Curr. Pollut. Rep. **6**(4), 399–409 (2020). https://doi.org/10.1007/s40726-020-00159-z

Liu, X., Lu, D., Zhang, A., Liu, Q., Jiang, G.: Data-driven machine learning in environmental pollution: gains and problems. Environ. Sci. Technol. **56**(4), 2124–2133 (2022). https://doi.org/ 10.1021/acs.est.1c06157

Long, M.S., Yantosca, R., Nielsen, J.E., Keller, C.A., da Silva, A., Sulprizio, M.P., Pawson, S., Jacob, D.J.: Development of a grid-independent GEOS-Chem chemical transport model (v9–02) as an atmospheric chemistry module for Earth system models. Geoscientific Model Develop. **8**, 595–602 (2015). https://doi.org/10.5194/gmd-8-595-2015

Mathur, R., Xing, J., Gilliam, R., Sarwar, G., Hogrefe, C., Pleim, J., Pouliot, G., Roselle, S., Spero, T.L., Wong, D.C., Young, J.: Extending the Community Multiscale Air Quality (CMAQ) modeling system to hemispheric scales: overview of process considerations and initial applications. Atmos. Chem. Phys. **17**, 12449–12474 (2017). https://doi.org/10.5194/acp-17-12449-2017

Minh, V.T.T., Tin, T.T. and Hien, T.T.: $PM_{2.5}$ forecast system by sing machine learning and WRF model, a case study: Ho Chi Minh City, Vietnam. Aerosol Air Qual. Res. **21**(12), 210108 (2021). https://doi.org/10.4209/aaqr.210108

Moosavi, V., Aschwanden, G., Velasco, E.: Finding candidate locations for aerosol pollution monitoring at street level using a data-driven methodology. Atmos. Measure. Tech. **8**, 3563–3575 (2015). https://doi.org/10.5194/amt-8-3563-2015

Nguyen, G.T.H., Shimadera, H., Uranishi, K., Matsuo, T., Kondo, A.: Numerical assessment of $PM_{2.5}$ and O_3 air quality in Continental Southeast Asia: Impacts of potential future climate change. Atmos. Environ. **215**, 116901 (2019). https://doi.org/10.1016/j.atmosenv.2019.116901

Peuch, V., Engelen, R., Rixen, M., Dee, D., Flemming, J., Suttie, M., Ades, M., Agustí-Panareda, A., Ananasso, C., Andersson, E., Armstrong, D., Barré, J., Bousserez, N., Dominguez, J. J., Garrigues, S., Inness, A., Jones, L., Kipling, Z., Letertre-Danczak, J., Parrington, M., Razinger, M., Ribas, R., Vermoote, S., Yang, X., Simmons, A., Garcés de Marcilla, J., Thépaut, J.: The Copernicus Atmosphere Monitoring Service: from research to operations. Bull. Am. Meteorol. Soc. **103**(12), E2650-E2668 (2022). https://doi.org/10.1175/BAMS-D-21-0314.1

Rakholia, R., Le, Q., Vu, K., Ho, B.Q., Carbajo, R.S.: AI-based air quality $PM_{2.5}$ forecasting models for developing countries: a case study of Ho Chi Minh City, Vietnam. Urban Clim. **46**, 101315 (2022). https://doi.org/10.1016/j.uclim.2022.101315

Reddington, C.L., Conibear, L., Robinson, S., Knote, C., Arnold, S.R., Spracklen, D.V.: Air pollution from forest and vegetation fires in Southeast Asia disproportionately impacts the poor. GeoHealth **5**(9), e2021GH000418 (2021). https://doi.org/10.1029/2021GH000418

Reddington, C.L., Yoshioka, M., Balasubramanian, R., Ridley, D., Toh, Y.Y., Arnold, S.R., Spracklen, D.V.: Contribution of vegetation and peat fires to particulate air pollution in Southeast Asia. Environ. Res. Lett. **9**, 094006 (2014). https://doi.org/10.1088/1748-9326/9/9/094006

Rémy, S., Kipling, Z., Huijnen, V., Flemming, J., Nabat, P., Michou, M., Ades, M., Engelen, R., Peuch, V.H.: Description and evaluation of the tropospheric aerosol scheme in the Integrated Forecasting System (IFS-AER, cycle 47R1) of ECMWF. Geoscientific Model Develop. **15**(12), 4881–4912 (2022). https://doi.org/10.5194/gmd-15-4881-2022

Sofiev, M., Vira, J., Kouznetsov, R., Prank, M., Soares, J., Genikhovich, E.: Construction of the SILAM Eulerian atmospheric dispersion model based on the advection algorithm of Michael Galperin. Geoscientific Model Develop. **8**, 3497–3522 (2015). https://doi.org/10.5194/gmd-8-3497-2015

Solazzo, E., Bianconi, R., Vautard, R., Appel, K.W., Moran, M.D., Hogrefe, C., Bessagnet, B., Brandt, J., Christensen, J.H., Chemel, C., Coll, I., Denier van der Gon, H., Ferreira, J., Forkel, R., Francis, X.V., Grell, G., Grossi, P., Hansen, A.B., Jeričević, A., Kraljević, L., Miranda, A.I., Nopmongcol, U., Pirovano, G., Prank, M., Riccio, A., Sartelet, K.N., Schaap, M., Silver, J.D., Sokhi, R.S., Vira, J., Werhahn, J., Wolke, R., Yarwood, G., Zhang, J., Rao, S.T., Galmarini, S.: Model evaluation and ensemble modelling of surface-level ozone in Europe and North America in the context of AQMEII. Atmos. Environ. **53**, 60–74 (2012). https://doi.org/10.1016/j.atmosenv.2012.01.003

Stein, A.F., Draxler, R.R., Rolph, G.D., Stunder, B.J., Cohen, M.D., Ngan, F.: NOAA's HYSPLIT atmospheric transport and dispersion modeling system. Bull. Am. Meteor. Soc. **96**(12), 2059–2077 (2015). https://doi.org/10.1175/BAMS-D-14-00110.1

Thongthammachart, T., Shimadera, H., Araki, S., Matsuo, T., Kondo, A.: Land use regression model established using light gradient boosting machine incorporating the WRF/CMAQ model for highly accurate spatiotemporal $PM_{2.5}$ estimation in the central region of Thailand. Atmos. Environ. **297**, 119595 (2023). https://doi.org/10.1016/j.atmosenv.2023.119595

Velasco, E., Rastan, S.: Air quality in Singapore during the 2013 smoke-haze episode over the Strait of Malacca: Lessons learned. Sustain. Cities Soc. **17**, 122–131 (2015). https://doi.org/10.1016/j.scs.2015.04.006

Vongruang, P., Pimonsree, S.: Biomass burning sources and their contributions to PM_{10} concentrations over countries in mainland Southeast Asia during a smog episode. Atmos. Environ. **228**, 117414 (2020). https://doi.org/10.1016/j.atmosenv.2020.117414

Wang, X., Fu, T.M., Zhang, L., Lu, X., Liu, X., Amnuaylojaroen, T., Latif, M.T., Ma, Y., Zhang, L., Feng, Zhu, L., Shen, H., Yang, X.: Rapidly changing emissions drove substantial surface and tropospheric ozone increases over Southeast Asia. Geophys. Res. Lett. **49**(19), e2022GL100223 (2022). https://doi.org/10.1029/2022GL100223

World Meteorological Organization: Training Materials and Best Practices for Chemical Weather/Air Quality Forecasting. Collection and series: ETR-No. 26, WMO Publications Board, Genova, Switzerland. https://library.wmo.int/doc_num.php?explnum_id=10439 (2020)

Xian, P., Reid, J.S., Hyer, E.J., Sampson, C.R., Rubin, J.I., Ades, M., Asencio, N., Basart, S., Benedetti, A., Bhattacharjee, P.S., Brooks, M.E., Colarco, P.R., da Saliva, A.M., Eck, T.F., Guth, J., Jorba, O., Kouznetsov, R., Kipling, Z., Sofiev, M., Garcia-Pando, C.P., Pradhan, Y., Tanaka, T., Wang, J., Westphal, D.L., Yumioto, K., Zhang, J.: Current state of the global operational aerosol multi-model ensemble: an update from the International Cooperative for Aerosol Prediction (ICAP). Q. J. R. Meteorol. Soc. **145**(51), 176–209 (2019). https://doi.org/10.1002/qj.3497

Chapter 8
Satellite-Based Remote Sensing

Abstract Advances in satellite remote sensing technology now enable observations of the atmosphere with unprecedented spatial and temporal resolution, allowing satellites to be used as a complement to ground level air quality monitoring. Nonetheless, the integration of satellite observations into air quality management in Southeast Asia is in its early stages. This situation is projected to change as GEMS, the first geostationary satellite mission designed to support air quality management in Asia, already provides continuous monitoring of a suite of atmospheric pollutants at fine spatial resolution across nearly the entire region of Southeast Asia. This chapter discusses how satellite-based remote sensing enables the tracking of air pollution patterns in urban and suburban areas, the transport of plumes from large individual emission sources and biomass burning, the production of ozone and other secondary pollutants downwind, and the improvement of emission inventories and air quality forecasting. The studies that have used satellite products to examine the distribution of pollutants over the region or the impact of specific events are also reviewed.

Keywords Satellite remote sensing · Low earth orbit satellite · Geostationary satellite

Satellites have become important information sources of atmospheric data, initially in the context of meteorology and climate change, but more recently in the chemical composition of the atmosphere. Advances in satellite remote sensing technology now allow observations of the atmosphere with unprecedented spatial and temporal resolution, allowing satellites to be used as a complement to ground level air quality monitoring. Nowadays satellite data can be used to assess air pollution in urban and suburban areas and its impact downwind, pollution transport from biomass burning and dust storms, and ozone (O_3) production at regional scale. Satellite products can also be used to precisely locate where pollution originates, the movement of pollution plumes, and how emission sources change over time (Holloway et al. 2021; Anenberg et al. 2020; Duncan et al. 2014; Fishman et al. 2008).

The obvious advantages of acquiring measurements remotely are the coverage of large geographical areas, the provision of data in otherwise remote or inaccessible areas, and the acquisition of measurements with consistency in time and space. Until recently, air quality satellite observations were made from sun-synchronous low Earth orbit (LEO) satellites, which pass over the Earth's poles once or twice a day. Because of their low temporal resolution, these satellites cannot track variations in air pollution throughout the diurnal course, nor can they track the progression of specific episodes, such as smoke-haze events triggered by wildfires. Fortunately, the advent of geostationary satellites has resolved this impediment, and it is now feasible to conduct repeated daytime observations of a given region. Geostationary satellites orbit synchronous with the Earth's rotation, allowing them to provide hourly data over a defined region during daylight hours. However, both LEO and geostationary satellites have the drawback of measuring the amount of atmospheric constituents along a column that crosses the troposphere and stratosphere, and therefore can only provide indirect estimates of pollutants abundance at surface level after going through assimilation routines in which satellite observations are combined with *in-situ* measurements at ground level and numerical model outputs.

In the case of aerosols, satellite column observations are used to calculate the aerosol optical depth (AOD). In the case of trace gases, satellite observations represent the density of molecules per unit area throughout the atmosphere's column and are reported in units of vertical column density (VCD, molecules cm^{-2}). Ozone's VCD can be reported in Dobson units (DU), a measure of 'thickness' where one DU is equivalent to 0.01 mm layer of the gas over a given area at standard temperature and pressure. The AOD and trace gases VCD are obtained after processing the original satellite readings (i.e., electromagnetic radiation signals) by applying so-called retrieval algorithms. The application of retrieval algorithms is a major challenge in Southeast Asia due to frequent cloudiness and a lack of information on the properties of the particles to calculate the AOD. Similarly, the limited number of reference-grade monitors at ground level jeopardizes the assimilation of satellite products to derive representative and accurate concentrations of trace gases at the surface.

The majority of instruments used to detect trace gases and aerosols are mounted on LEO satellites. Only two geostationary satellites equipped to measure atmospheric pollutants have been put into operation, one in Asia and the other in North America, with the former covering most of Southeast Asia. Many LEO satellites are part of satellite trains, which are suites of satellites that orbit in close proximity to one another. Satellite trains enable near-simultaneous coordinated measurements that can be used together to obtain comprehensive information about atmospheric components or processes that occur at the same time. One example is NASA's A-Train, which consists of four satellites, three of which (Terra, Aura and Aqua) provide useful information for air quality management. Martin (2008) and L'Ecuyer and Jiang (2010) provide detailed descriptions of LEO satellites and satellite trains, respectively.

Table 8.1 lists the major satellite instruments designed for remote sensing of aerosols and chemically reactive trace gases in the lower troposphere. Only instruments that were in orbit at the time of writing are included. Solar backscattering instruments are only used during the day to measure nitrogen dioxide (NO_2), formaldehyde (HCHO), glyoxal (CHOCHO), and AOD. While thermal infrared instruments are used to measure carbon monoxide (CO) and sulfur dioxide (SO_2)

during the day and at night. Both types of instruments can be used to measure O_3. Most instruments achieve global coverage in the absence of clouds on a timescale of days. It is worth noting that the spatial resolution has improved over the years.

Satellite data and products are made publicly available by space agencies through their web portals or central servers. Users must take into account the level of processing of the data available at the time of download. Space agencies and independent organizations provide resources to ingest, process, analyze, and visualize remote sensing data. Examples of these resources are the Earth Observations (https://neo.gsfc.nasa.gov/), Worldview (https://worldview.Earthdata.nasa.gov) and Giovanni (https://giovanni.gsfc.nasa.gov/giovanni/) applications managed by NASA, and the Copernicus Data Space Ecosystem (https://dataspace.copernicus.eu/) and the Atmospheric Toolbox (https://atmospherictoolbox.org/) managed by the European Union Space Agency and the Copernicus Program.

8.1 Geostationary Satellites

There are currently three geostationary satellites that provide observations related to air quality for Southeast Asia. Since 2015, the Himawari-8 and -9 satellites managed by the Japan Meteorology Association have provided AOD observations every 10 min, while the Republic of Korea's GEO-KOMPSAT 2B satellite launched in 2020 provides observations of AOD and a number of reactive trace gases relevant to air quality management.

The Geostationary Environment Monitoring Spectrometer (GEMS) onboard GEO-KOMPSAT 2B is the first instrument to enable continuous monitoring of a suite of atmospheric pollutants at fine spatial resolution over a specific region of the world. Rather than tracking air pollution, the Himawari-8 and -9 satellites are intended for meteorological purposes. GEMS provides hourly observations eight times during daylight hours across East Asia, a portion of Central Asia, and Southeast Asia, covering the entire mainland subregion as well as much of the maritime subregion. For the region in question, Brunei Darussalam, Cambodia, Lao PDR, Malaysia, Myanmar, Philippines, Singapore, Thailand, and Vietnam are fully covered, Indonesia is partially covered, while Timor-Leste is left out. Himawari-8 and -9 cover the entire Southeast Asian region.

GEMS is a step-and-stare scanning spectrometer sensitive to visible and ultraviolet wavelengths, with nominal spatial resolutions of 7 km × 8 km and 3.5 km × 8 km per pixel for gases and aerosols, respectively. It has a spectral range of 300–500 nm with a spectral resolution of 0.6 nm and provides spectra of O_3, NO_2, SO_2, HCHO, CHOCHO, and other parameters associated with atmospheric chemistry cycles (Kim et al. 2020). GEMS, in conjunction with retrieval algorithms, is capable of resolving the abundance of pollutants in the low troposphere with sufficient spatial resolution and frequency to address critical inquiries regarding air quality. GEMS is one of three air quality monitoring missions over the northern hemisphere; the other two are NASA's Tropospheric Emissions: Monitoring of Pollution (TEMPO) and ESA's

Table 8.1 Major active satellite instruments covering Southeast Asia designed for remote sensing of aerosols and chemically reactive trace gases in the lower troposphere

Mission/Instrument	Satellite	Type of satellite[a]	Entered service	Pollutant species	Nominal spatial resolution (km)	Temporal resolution	Data repository/Instrument website
Measurements Of Pollution In The Troposphere (MOPPITT)	Terra[b]	LEO	2000	CO	22 × 22	3.5 days	https://www2.acom.ucar.edu/mopitt
Multiangle Imaging Spectro Radiometer (MISR)	Terra[b]	LEO	2000	AOD	18 × 18	7 days	https://misr.jpl.nasa.gov/
Moderate Resolution Imaging Spectroradiometer (MODIS)	Terra[b] & Aqua[c]	LEO	2000 and 2002	AOD	10 × 10	2 days	https://modis.gsfc.nasa.gov/
Atmospheric Infrared Sounder (AIRS)	Aqua[c]	LEO	2002	SO_2, CO	14 × 14	1 day	https://airs.jpl.nasa.gov/
Ozone Monitoring Instrument (OMI)	Aura[d]	LEO	2004	O_3, NO_2, SO_2, HCHO, BrO, OClO, AOD	24 × 13	1 day	https://disc.gsfc.nasa.gov/ https://www.knmiprojects.nl/projects/ozone-monitoring-instrument
Cloud-Aerosol Lidar and Infrared Pathfinder Satellite Observation (CALIPSO)	CALIPSO[e]	LEO	2006	AOD	40 × 40	30 days	https://www-calipso.larc.nasa.gov/ https://calipso.cnes.fr/en/CALIPSO/index.htm
Thermal and Near-Infrared Sensor for Carbon Observation—Fourier Transform Spectrometer (TANSO-FTS)	GOSAT[f]	LEO	2009	CO_2, CH_4	10.5 × 10.5	3 days	https://www.gosat.nies.go.jp/en/

(continued)

Table 8.1 (continued)

Mission/Instrument	Satellite	Type of satellite[a]	Entered service	Pollutant species	Nominal spatial resolution (km)	Temporal resolution	Data repository/Instrument website
Thermal and Near-Infrared Sensor for Carbon Observation—Cloud and Aerosol Imager (TANSO-CAI)	GOSAT[f]	LEO	2009	AOD	0.5×0.5	3 days	https://www.gosat.nies.go.jp/en/
Cross-track Infrared Sounder (CrIs)	S-NNP[g]	LEO	2011	O_3, CO, CH_4, NH_3, PAN	14×14	1 day	https://www.earthdata.nasa.gov/sensors/cris https://www.earthdata.nasa.gov/eosdis/daacs/gesdisc
Global Ozone Monitoring Experiment—2 (GOME-2)	MetOp-b[h]	LEO	2012	O_3, NO_2, SO_2, HCHO, BrO	80×40	1 day	https://atmos.caf.dlr.de/app/missions/gome2 https://acsaf.org/ https://www.eumetsat.int/gome-2
Infrared Atmospheric Sounding Interferometer (IASI)	MetOp-b[h]	LEO	2012	CO, CH_4, N_2O, NH_3	12×12	1 day	https://www.ospo.noaa.gov/Products/atmosphere/soundings/heap/iasi/iasiproducts.html https://www.eumetsat.int/iasi

(continued)

Table 8.1 (continued)

Mission/Instrument	Satellite	Type of satellite[a]	Entered service	Pollutant species	Nominal spatial resolution (km)	Temporal resolution	Data repository/ Instrument website
Advanced Himawari Imager (AHI)	Himawari-8 & -9[i]	GEO	2015 & 2017	AOD	2×2	10 min	https://www.eorc.jaxa.jp/ ptree/index.html https://www.jma.go.jp/ jma/jma-eng/satellite/ index.html
TROPOspheric Monitoring Instrument (TROPOMI)	Sentinel-5P[j]	LEO	2017	O_3, NO_2, SO_2, CO, HCHO, CH_4, AOD	3.5×6	1 day	https://s5phub.copern icus.eu/dhus/#/home http://www.tropomi.eu/
Cross-track Infrared Sounder (CrIs)	NOAA-20/JPSS[g]	LEO	2017	O_3, CO, CH_4, NH_3, PAN	14×14	1 day	https://www.earthdata. nasa.gov/sensors/cris https://www.earthdata. nasa.gov/eosdis/daacs/ gesdisc
Thermal and Near-Infrared Sensor for Carbon Observation—Fourier Transform Spectrometer 2 (TANSO-FTS-2)	GOSAT-2 [k]	LEO	2018	CO_2, CH_4, O_3, CO	1×1	6 days	https://www.gosat-2.nies. go.jp/

(continued)

Table 8.1 (continued)

Mission/Instrument	Satellite	Type of satellite[a]	Entered service	Pollutant species	Nominal spatial resolution (km)	Temporal resolution	Data repository/Instrument website
Thermal and Near-Infrared Sensor for Carbon Observation—Cloud and Aerosol Imager 2 (TANSO-CAI-2)	GOSAT-2[k]	LEO	2018	AOD	1×1	6 days	https://www.gosat-2.nies.go.jp/
Geostationary Environment Monitoring Spectrometer (GEMS)	GEO-KOMPSAT 2B[l]	GEO	2020	O_3, NO_2, SO_2, HCHO, CHOCHO, AOD	Gases: 7×8, aerosols 3.5×8	1 h	https://nesc.nier.go.kr/

[a] LEO and GEO stand for low Earth orbit and geostationary satellites, respectively

[b] Terra decommissioning is expected to occur around 2025–26 (https://terra.nasa.gov/)

[c] Aqua is expected to operate until August 2026 when power generation will end (https://aqua.nasa.gov/)

[d] Aura is expected to operate beyond 2023 as far as 2036 (https://aura.gsfc.nasa.gov/index.html)

[e] Calipso is a joint US's NASA and France's Centre National d'Etudes Spatiales (CNES) satellite mission expecting to operate until the end of 2023 (https://calipso.cnes.fr/en/CALIPSO/index.htm)

[f] The Greenhouse gases Observing SATellite (GOSAT) also known as *Ibuki* (breath in Japanese) was planned for a 5-year mission, but it is still in operation (https://www.gosat.nies.go.jp/en/)

[g] The Suomi-National Polar-orbiting Partnership (S-NPP) satellite and the National Oceanic and Atmospheric Administration NOAA-20 satellite, also known as JPSS-1, are part of the Joint Polar Satellite System (JPSS) (https://www.nesdis.noaa.gov/current-satellite-missions/currently-flying/joint-polar-satellite-system)

[h] MetOp-b is expected to operate at least until 2027 (https://www.eumetsat.int/our-satellites/metop-series)

[i] Himawari-9 has currently the role of main observer, while Himawari-8 has a backup role. Both satellites are scheduled to operate until 2029 (https://www.jma.go.jp/jma/jma-eng/satellite/index.html)

[j] Sentinel-5P mission is planned to operate at least until 2024 (https://www.esa.int/Applications/Observing_the_Earth/Copernicus/Sentinel-5P)

[k] The Greenhouse gases Observing SATellite 2 (GOSAT-2) is the successor of GOSAT and was originally planned for a 5-year mission (https://www.gosat-2.nies.go.jp/)

[l] The Geostationary Korea Multi-Purpose Satellite 2B (GEO-KOMPSAT 2B) also known as Cheollian 2B is designed to operate at least 10 years (https://nesc.nier.go.kr/)

Sentinel-4, which are commissioned to monitor air pollution across North America and Europe, respectively. The following are the objectives of the GEMS mission:

- Provide high temporal and spatial resolution atmospheric chemistry measurements over Asia.
- Monitor transboundary haze pollution and Asian dust.
- Improve understanding of the interactions between atmospheric chemistry and meteorology.
- Better understand the global aspects of tropospheric pollution.
- Improve air quality forecasting systems by constraining emission rates and assimilating data from chemical observations.

Figure 8.1 shows examples of GEMS products related to air quality. GEMS had been in operation for nearly five years at the time of writing, and Korea's Environmental Satellite Center was still working on validating its retrieval algorithms. However, GEMS products were already being used in a number of studies (see special issue 'GEMS: first year in operation' jointly organized by the journals Atmospheric Measurement Techniques and Atmospheric Chemistry and Physics). Importantly, GEMS is the first instrument to observe air quality from a geostationary earth orbit; consequently, the development of its products is still ongoing.

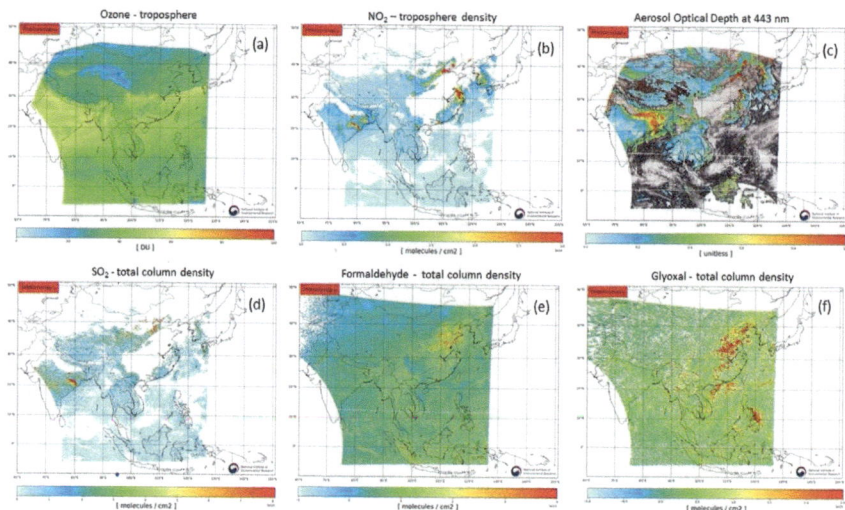

Fig. 8.1 Examples of GEMS products related to air quality: amount of O_3 in the troposphere expressed in Dobson units (**a**), density of NO_2 per unit area in the troposphere (**b**), aerosol optical depth (**c**), and densities of SO_2, HCHO, and CHOCHO from the surface to the top of the atmosphere per unit area (**d–f**). The retrievals correspond to 20 December 2022 at 4:45 h UTC (13:45 Korea local time). Note that all products are labeled as preliminary. At the time of writing, scientists at Korea's Environmental Satellite Center were still working on validating the retrieval algorithms. Satellite data and images are available on the GEMS web portal: https://nesc.nier.go.kr/

8.2 Use of Satellite Products Related to Air Quality in Southeast Asia

Although satellite data on air quality is becoming more widely available, it has not yet been fully integrated into air quality management in Southeast Asia. However, all of the forecast systems and some of the air quality modeling studies discussed in previous chapters have used satellite products to infer, simulate, and validate air pollution loading across the globe and the region (see Table 7.1). The Copernicus Atmospheric Monitoring Service (CAMS), for example, assimilates over 90 different satellite meteorological and atmospheric composition data streams, including data from nine instruments listed in Table 8.1 (Peuch et al. 2022). Similarly, the Goddard Earth Observing Composition Forecasting (GEOS-CF) system uses data from NASA's Earth Observation missions, including data from instruments mounted in the A-Train (Keller et al. 2021). Regarding the use of satellite products to validate outputs of air quality models in academic studies, some authors have used observations of AOD, total O_3 column, and tropospheric columns of NO_2 and CO to evaluate the spatial distribution of simulated aerosols and pollutant gases (e.g., Wang et al. 2022; Li et al. 2022; Vongruang and Pimonsree 2020; Crippa et al. 2016; Aouizerats et al. 2015; Dong and Fu 2015; Amnuaylojaroen et al. 2014; Fu et al. 2012).

Satellite products have also been used to assess the local and regional impacts of biomass burning and anthropogenic emissions in Southeast Asia (e.g., Ohno et al. 2022; Salvador et al. 2022; Napi et al. 2022; Kalita et al. 2020; Nguyen et al. 2019; Yin et al. 2019; Gautam et al. 2013; Deng et al. 2008). Likewise, during the COVID-19 pandemic, satellite data was used in Southeast Asia, as it was in the rest of the world, to assess the impacts of the lockdown on air quality (e.g., Wetchayont et al. 2021; Kanniah et al. 2020; Metya et al. 2020; Stratoulias and Nuthammachot 2020). Among the satellite-based studies on aerosol pollution, Banerjee et al. (2021) and Yin (2023a,b) stand out. The former study investigated the aerosol climatology over the region using products associated with aerosol loading and its optical and microphysical properties derived from NASA's A-Train. While the latter two studies integrated $PM_{2.5}$ ambient concentrations derived from validated satellite-based datasets, in which AOD retrievals from several satellite instruments were combined and related to surface $PM_{2.5}$ measurements and outputs of a chemical transport model, with the Global Exposure Mortality Model to estimate the number of premature deaths in Southeast Asia over the past 30 years. Regarding other pollutant species, it is worth highlighting the study carried out by Gaudel et al. (2024). They discovered that Southeast Asia, specifically Malaysia and Indonesia, is the region in the tropics where tropospheric O_3 has increased the most over the last 25 years, due in part to increased ozone precursor emissions.

References

Amnuaylojaroen, T., Barth, M.C., Emmons, L.K., Carmichael, G.R., Kreasuwun, J., Prasitwat-tanaseree, S., Chantara, S.: Effect of different emission inventories on modeled ozone and carbon monoxide in Southeast Asia. Atmos. Chem. Phys. **14**, 12983–13012 (2014). https://doi.org/10.5194/acp-14-12983-2014

Anenberg, S.C., Bindl, M., Brauer, M., Castillo, J.J., Cavalieri, S., Duncan, B.N., Fiore, A.M., Fuller, R., Goldberg, D.L., Henze, D.K. and Hess, J., Halloway, T., James, P., Jin, X., Kheirbek. I., Kinney, P.L., Mohegn, A., Patz, J., Jimenez, M.P., Rot, A., Tong, D., Walker, K., Watts, N., West, J.: Using satellites to track indicators of global air pollution and climate change impacts: Lessons learned from a NASA-supported science-stakeholder collaborative. GeoHealth **4**(7), e2020GH000270 (2020). https://doi.org/10.1029/2020GH000270

Aouizerats, B., van der Werf, G.R., Balasubramanian, R., Betha, R.: Importance of transboundary transport of biomass burning emissions to regional air quality in Southeast Asia during a high fire event. Atmos. Chem. Phys. **15**, 363–373 (2015). https://doi.org/10.5194/acp-15-363-2015

Banerjee, T., Shitole, A.S., Mhawish, A., Anand, A., Ranjan, R., Khan, M.F., Srithawirat, T., Latif, M.T., Mall, R.K.: Aerosol climatology over South and Southeast Asia: aerosol types, vertical profile, and source fields. J. Geophys. Res. Atmos. **126**(6), e2020JD033554 (2021). https://doi.org/10.1029/2020JD033554

Crippa, P., Castruccio, S., Archer-Nicholls, S., Lebron, G.B., Kuwata, M., Thota, A., Sumin, S., Butt, E., Wiedinmyer, C., Spracklen, D.V.: Population exposure to hazardous air quality due to the 2015 fires in Equatorial Asia. Sci. Rep. **6**, 37074 (2016). https://doi.org/10.1038/srep37074

Deng, X., Tie, X., Zhou, X., Wu, D., Zhong, L., Tan, H., Li, F., Huang, X., Bi, X., Deng, T.: Effects of Southeast Asia biomass burning on aerosols and ozone concentrations over the Pearl River Delta (PRD) region. Atmos. Environ. **42**(36), 8493–8501 (2008). https://doi.org/10.1016/j.atmosenv.2008.08.013

Dong, X., Fu, J.S.: Understanding interannual variations of biomass burning from Peninsular Southeast Asia, part I: model evaluation and analysis of systematic bias. Atmos. Environ. **116**, 293–307 (2015). https://doi.org/10.1016/j.atmosenv.2015.06.026

Duncan, B.N., Prados, A.I., Lamsal, L.N., Liu, Y., Streets, D.G., Gupta, P., Hilsenrath, E., Kahn, R.A., Nielsen, J.E., Beyersdorf, A.J., Burton, S.P., Fiore, A.M., Fishman, J., Henze, D.K., Hostetler, C.A., Krotkov N.A., Lee, P., in, M., Pawson, S., Pfister, G., Pickering, K.E., Pierce, R.B., Yoshida, Y., Ziemba, L.D.: Satellite data of atmospheric pollution for US air quality applications: Examples of applications, summary of data end-user resources, answers to FAQs, and common mistakes to avoid. Atmos. Environ. **94**, 647–662 (2014). https://doi.org/10.1016/j.atmosenv.2014.05.061

Fishman, J., Bowman, K.W., Burrows, J.P., Richter, A., Chance, K.V., Edwards, D.P., Martin, R.V., Morris, G.A., Pierce, R.B., Ziemke, J.R., Al-Saadi, J.A., Creilson, J.K., Schack, T.K., Thompson, A.M.: Remote sensing of tropospheric pollution from space. Bull. Am. Meteor. Soc. **89**(6), 805–822 (2008). https://doi.org/10.1175/2008BAMS2526.1

Fu, J.S., Hsu, N.C., Gao, Y., Huang, K., Li, C., Lin, N.-H., Tsay, S.-C.: Evaluating the influences of biomass burning during 2006 BASE-ASIA: a regional chemical transport modeling. Atmos. Chem. Phys. **12**, 3837–3855 (2012). https://doi.org/10.5194/acp-12-3837-2012

Gaudel, A., Bourgeois, I., Li, M., Chang, K.-L., Ziemke, J., Sauvage, B., Stauffer, R.M., Thompson, A.M., Kollonige, D.E., Smith, N., Hubert, D., Keppens, A., Cuesta, J., Heue, K.-P., Veefkind, P., Aikin, K., Peischl, J., Thompson, C.R., Ryerson, T.B., Frost, G.J., McDonald, B.C., Cooper, O.R.: Tropical tropospheric ozone distribution and trends from in situ and satellite data. Atmos. Chem. Phys. **24**, 9975–10000 (2024). https://doi.org/10.5194/acp-24-9975-2024

Gautam, R., Hsu, N.C., Eck, T.F., Holben, B.N., Janjai, S., Jantarach, T., Tsay, S.C., Lau, W.K.: Characterization of aerosols over the Indochina peninsula from satellite-surface observations during biomass burning pre-monsoon season. Atmos. Environ. **78**, 51–59 (2013). https://doi.org/10.1016/j.atmosenv.2012.05.038

Holloway, T., Miller, D., Anenberg, S., Diao, M., Duncan, B., Fiore, A.M., Henze, D.K., Hess, J., Kinney, P.L., Liu, Y., Neu, J.L., O'Neil, S.M., Odman, M.T., Pierce, R.B., Russell, A.G., Tong, D., West, J.J., Zondlo, M.A.: Satellite monitoring for air quality and health. Annu. Rev. Biomed. Data Sci. **4**, 417–447 (2021). https://doi.org/10.1146/annurev-biodatasci-110920-093120

Kalita, G., Kunchala, R.K., Fadnavis, S., Kaskaoutis, D.G.: Long term variability of carbonaceous aerosols over Southeast Asia via reanalysis: association with changes in vegetation cover and biomass burning. Atmos. Res. **245**, 105064 (2020). https://doi.org/10.1016/j.atmosres.2020.105064

Kanniah, K.D., Zaman, N.A.F.K., Kaskaoutis, D.G., Latif, M.T.: COVID-19's impact on the atmospheric environment in the Southeast Asia region. Sci. Total Environ. **736**, 139658 (2020). https://doi.org/10.1016/j.scitotenv.2020.139658

Keller, C.A., Knowland, K.E., Duncan, B.N., Liu, J., Anderson, D.C., Das, S., Lucchesi, R.A., Lundgren, E.W., Nicely, J.M., Nielsen, E., Ott, L.E., Saunders, E., Strode, S.A., Wales, P. A., Jacob. D. J., Pawson, S.: Description of the NASA GEOS composition forecast modeling system GEOS-CF v1. 0. J. Adv. Model. Earth Syst. **13**(4), e2020MS002413 (2021). https://doi.org/10.1029/2020MS002413

Kim, J., Jeong, U., Ahn, M., Kim, J.H., Park, R.J., Lee, H., Song, C.H., Choi, Y., Lee, K., Yoo, J., Jeong, M., Park, S.K., Lee, K., Song, C., Kim, S., Kim, Y.J., Kim, S., Kim, M., Go, S., Liu, X., Chance, K., Chan Miller, C., Al-Saadi, J., Veihelmann, B., Bhartia, P.K., Torres, O., Abad, G.G., Haffner, D.P., Ko, D.H., Lee, S.H., Woo, J., Chong, H., Park, S.S., Nicks, D., Choi, W.J., Moon, K., Cho, A., Yoon, J., Kim, S., Hong, H., Lee, K., Lee, H., Lee, S., Choi, M., Veefkind, P., Levelt, P.F., Edwards, D.P., Kang, M., Eo, M., Bak, J., Baek, K., Kwon, H., Yang, J., Park, J., Han, K.M., Kim, B., Shin, H., Choi, H., Lee, E., Chong, J., Cha, Y., Koo, J., Irie, H., Hayashida, S., Kasai, Y., Kanaya, Y., Liu, C., Lin, J., Crawford, J.H., Carmichael, G.R., Newchurch, M.J., Lefer, B.L., Herman, J.R., Swap, R.J., Lau, A.K.H., Kurosu, T.P., Jaross, G., Ahlers, B., Dobber, M., McElroy, C.T., Choi, Y.: New era of air quality monitoring from space: Geostationary Environment Monitoring Spectrometer (GEMS). Bull. Am. Meteorol. Soc. **101**(1), E1–E22 (2020). https://journals.ametsoc.org/view/journals/bams/101/1/bams-d-18-0013.1.xml

L'Ecuyer, T.S., Jiang, J.H.: Touring the atmosphere aboard the A-Train. Phys. Today **63**(7), 36–41 (2010). https://doi.org/10.1063/1.3463626

Li, J., Han, Z., Surapipith, V., Fan, W., Thingboonchoo, N., Wu, J., Li, J., Tao, J., Wu, Y., Macatangay, R., Bran, S.H., Yu, E., Zhang, A., Liang, L., Zhang, R.: Direct and indirect effects and feedbacks of biomass burning aerosols over Mainland Southeast Asia and South China in springtime. Sci. Total Environ. **842**, 156949 (2022). https://doi.org/10.1016/j.scitotenv.2022.156949

Martin, R.V.: Satellite remote sensing of surface air quality. Atmos. Environ. **42**(34), 7823–7843 (2008). https://doi.org/10.1016/j.atmosenv.2008.07.018

Metya, A., Dagupta, P., Halder, S., Chakraborty, S., Tiwari, Y.K.: COVID-19 lockdowns improve air quality in the South-East Asian regions, as seen by the remote sensing satellites. Aerosol Air Qual. Res. **20**(8), 1772–1782 (2020). https://doi.org/10.4209/aaqr.2020.05.0240

Napi, N.N.L.M., Ooi, M.C.G., Talif, M.L., Juneng, L., Nadzir, M.S.M., Chan, A., Li, L., Abdullah, S.: Contribution of aerosol species to the 2019 smoke episodes over the East Coast of Peninsular Malaysia. Aerosol Air Qual. Res. **22**(7), 210393 (2022). https://doi.org/10.4209/aaqr.210393

Nguyen, T.T., Pham, H.V., Lasko, K., Bui, M.T., Laffly, D., Jourdan, A., Bui, H.Q.: Spatiotemporal analysis of ground and satellite-based aerosol for air quality assessment in the Southeast Asia region. Environ. Pollut. **255**, 113106 (2019). https://doi.org/10.1016/j.envpol.2019.113106

Ohno, T., Irie, H., Momoi, M., da Silva, A.M.: Quantitative evaluation of mixed biomass burning and anthropogenic aerosols over the Indochina Peninsula using MERRA-2 reanalysis products validated by sky radiometer and MAX-DOAS observations. Prog. Earth Planet Sci. **9**, 61 (2022). https://doi.org/10.1186/s40645-022-00520-4

Peuch, V., Engelen, R., Rixen, M., Dee, D., Flemming, J., Ades, M., Agustí-Panareda, A., Ananasso, C., Andersson, E., Armstrong, D., Barré, J., Bousserez, N., Dominguez, J.J., Garrigues, S., Inness, A., Jones, L., Kipling, Z., Letertre-Danczak, J., Parrington, M., Razinger, M., Ribas, R., Vermoote, S., Yang, X., Simmons, A., Garcés de Marcilla, J., Thépaut, J.: The

Copernicus Atmosphere Monitoring Service: from research to operations. Bull. Am. Meteorol. Soc. **103**(12), E2650–E2668 (2022). https://doi.org/10.1175/BAMS-D-21-0314.1

Salvador, C.M.G., Coronel, I.C.V., VII, A.T.B., Sugcang, R.J., Lavapie, M.A.M., Capangpangan, R.Y., Pabroa, P.C.B.: Variability and source characterization of regional PM of two urban areas dominated by biomass burning and anthropogenic emission. Aerosol Air Qual. Res. **22**, 220026 (2022). https://doi.org/10.4209/aaqr.220026

Stratoulias, D., Nuthammachot, N.: Air quality development during the COVID-19 pandemic over a medium-sized urban area in Thailand. Sci. Total Environ. **746**, 141320 (2020). https://doi.org/10.1016/j.scitotenv.2020.141320

Vongruang, P., Pimonsree, S.: Biomass burning sources and their contributions to PM_{10} concentrations over countries in mainland Southeast Asia during a smog episode. Atmos. Environ. **228**, 117414 (2020). https://doi.org/10.1016/j.atmosenv.2020.117414

Wang, X., Fu, T.M., Zhang, L., Lu, X., Liu, X., Amnuaylojaroen, T., Latif, M.T., Ma, Y., Zhang, L., Feng, Zhu, L., Shen, H., Yang, X.: Rapidly changing emissions drove substantial surface and tropospheric ozone increases over Southeast Asia. Geophys. Res. Lett. **49**(19), e2022GL100223 (2022). https://doi.org/10.1029/2022GL100223

Wetchayont, P., Hayasaka, T., Khatri, P.: Air quality improvement during COVID-19 lockdown in Bangkok Metropolitan, Thailand: Effect of the long-range transport of air pollutants. Aerosol Air Qual. Res. **21**(7), 200662 (2021). https://doi.org/10.4209/aaqr.200662

Yin, S., Wang, X., Zhang, X., Guo, M., Miura, M., Xiao, Y.: Influence of biomass burning on local air pollution in mainland Southeast Asia from 2001 to 2016. Environ. Pollut. **254**(A), 112949 (2019). https://doi.org/10.1016/j.envpol.2019.07.117

Yin, S.: Decadal changes in premature mortality associated with exposure to outdoor $PM_{2.5}$ in mainland Southeast Asia and the impacts of biomass burning and anthropogenic emissions. Sci. Total Environ. **854**, 158775 (2023a). https://doi.org/10.1016/j.scitotenv.2022.158775

Yin, S.: Effect of biomass burning on premature mortality associated with long-term exposure to $PM_{2.5}$ in Equatorial Asia. J Environ. Manage. **330**, 117154 (2023b). https://doi.org/10.1016/j.jenvman.2022.117154

Chapter 9
Scientific Research

Abstract Scientific research is at the center of air quality management. Each of its components must be supported by scientific research to provide solid foundations for informed action. This chapter reviews the scientific achievements, challenges, needs, and gaps in the field throughout Southeast Asia. Efforts have focused on understanding the patterns, trends, and impact of particle pollution caused by biomass burning. Several studies have investigated the particles' chemical composition and physical properties, as well as their origin and effects on public health. A growing number of studies have also focused on other aspects of air quality in large cities such as Manila, Bangkok, Hanoi, Ho Chi Minh City, Kuala Lumpur, and Singapore, but little research has been undertaken outside them. Many relevant topics have received little or no attention, including atmospheric chemistry, climate change's impacts on air quality and vice versa, boundary layer meteorology, emergent and persistent pollutants, toxicology, and ecosystem damage, among others. Similarly, there is a need to develop monitoring and analysis methods that are tailored to Southeast Asia, as well as chemical-transport models and air quality multisystem analysis to download regional air quality forecasts and satellite products that can be used for air quality regulatory and warning purposes.

Keywords Particle characterization · Volatile organic compounds · Emissions characterization · Atmospheric chemistry · Personal exposure · Epidemiology and toxicology

Scientific research is critical in assisting environmental authorities in determining the pollutants' origin, transport and transformation in the atmosphere, and impact on human health and the environment. To ensure the successful implementation of air quality programs, scientific research is also required to identify effective emission control measures and monitor the development of existing regulations. Science contributions arise from monitoring (ambient concentrations, emissions, meteorology, and health exposure), analysis (modeling, cost-benefits, and risk assessment), research (atmospheric chemistry and physics, toxicology, and ecosystem damage), and development (monitoring and analysis methods, instrumentation, modeling systems, satellite retrieval algorithms and assimilation systems, and source control technologies). Scientific information provides fundamental knowledge to air quality managers, allowing them to make informed decisions.

As noted in previous chapters, the academic community has taken on the task of compiling emission inventories and running air quality models in Southeast Asia (see Tables 6.1 and 7.1). Similarly, researchers from local universities as well as overseas institutions are the ones who have made significant scientific contributions to the understanding and solution of many of the region's air quality problems. Table 9.1 presents a selection of 100 peer-reviewed articles published over the last decade that the authors consider representative of the region's efforts to better understand the processes driving air pollution. This list of articles should be used as a starting point when looking for scientific information on air pollution in the region. The table is divided into seven sections to accommodate the topics that have garnered the greatest interest among members of the scientific community. Some of the topics that were mentioned in the previous paragraph are not included because they have yet to be studied in the region. The table does not include the references that are listed in Tables 6.1 and 7.1, it only includes references that have not been discussed previously.

Following a careful review of the topics and content of the articles listed in Table 9.1, the following observations emerge:

- Much of the research has focused on the chemical characterization of particles as well as the study of their microphysical, optical, and radiative properties. This responds to the fact that the most critical air quality episodes in the region are typically caused by high levels of particle pollution, which poses a major threat to public health, in addition to the implications of particles in global climate forcing.
- Several studies have been conducted to assess the impact of biomass burning on particle loading, chemical composition, and seasonal variations. Studies in the mainland region have focused on the impact of burning agricultural residue (e.g., Thepnuan et al. 2019; Pani et al. 2018; Wang et al. 2015), whereas studies in the maritime region have focused on the impact of peatland fires (e.g., Tham et al. 2019; Budisulistiorini et al. 2018; Sulong et al. 2017). Adam et al. (2021) provide an extensive list of recent literature on the subject. They reviewed and summarized existing knowledge on particle pollution from biomass burning in the region.
- Similarly, significant efforts have been made to characterize the emissions from biomass burning. A number of field measurements and laboratory experiments have been carried out to determine the emission factors and emission profiles of pollutant gases and particles. Studies in Thailand and Vietnam have focused on characterizing emissions from burning rice straw and maize residues (e.g., Phuong et al. 2022; Pham et al. 2021; Chantara et al. 2019), while smoldering peat fires have been studied in Indonesia (e.g., Yokelson et al. 2022; Stockwell et al. 2016; Fujii et al. 2014).
- Aside from characterizing particulate pollution caused by biomass burning, little has been done on other regional air quality issues. The majority of air quality research has concentrated on seven cities: Manila, Hanoi, Ho Chi Minh City,

Table 9.1 Selected peer-reviewed literature on air quality research in Southeast Asia

Reference	Pollutant species/ parameters	Study area	Study period	Topic	Approach
Air pollution trends, seasonal variations, and correlations					
Sukkhum et al. (2022)	PM_{10}, SO_2, NO_X, NO_2, O_3, CO	Northern Thailand	2004–2018	Seasonal patterns and trends of criteria pollutants	Data from official air quality monitoring stations. Statistical analysis
Nghiem et al. (2021)	PM_{10}, $PM_{2.5}$, SO_2, NO, NO_2, O_3, CO	Hanoi, Vietnam	2010–2018	Probability distributions, trends, and temporal variations of criteria pollutants	Data from official air quality monitoring stations. Integrated approach to retrieve and analyze air quality data
Halim et al. (2020)	PM_{10}, SO_2, NO, NO_2, O_3, CO	Kuala Lumpur, Malaysia	2000–2015	Urbanization impact on air quality	Data from official air quality monitoring stations. Statistical and geostatistical analysis
Zulkepli et al. (2019)	PM_{10}	Kuala Lumpur and Jerantut, Malaysia	2000–2015	Topological characterization of smoke-haze episodes	Data from official air quality monitoring stations. Topological data analysis using the persistent homology method
Awang et al. (2018)	O_3, NO_2, NO, PM_{10}	Kuala Lumpur and Melaka, Malaysia	2013–2014	Ground-level O_3 concentrations during haze episodes	Data from official air quality monitoring stations. Statistical analysis and theoretical estimate of photolysis rates
Uttamang et al. (2018)	CO, SO_2, NO, NO_2, O_3	Bangkok, Thailand	2010–2014	Diurnal and seasonal variations, and interannual trends of gaseous criteria pollutants. Chemical and physical processes associated with O_3 episodes	Data from official air quality monitoring stations. Statistical analysis. Estimates of O_3 photolysis rate constant, and analyses to estimate local and regional O_3 contributions

(continued)

Table 9.1 (continued)

Reference	Pollutant species/parameters	Study area	Study period	Topic	Approach
Samsuddin et al. (2015)	PM_{10}, SO_2, NO, NO_2, O_3, CO	Johor and Kuala Lumpur, Malaysia	2015	Impacts of local and transboundary factors on criteria pollutants during the haze episode of 2015	Data from two official air quality monitoring stations. Statistical analysis. Cluster analysis of backward trajectories based on HYPLIST and wildfire radiative power data from the Global Fire Assimilation System (GFAS)
Dotse et al. (2016b)	PM_{10}	Brunei Darussalam	2009–2014	Temporal and spatial distribution, and potential emission sources of particle pollution across the country	Data from four official air quality monitoring stations. Statistical analysis based on bivariate concentration polar plots and k-means clusters. Use of HYSPLIT backward trajectories in combination with MODIS fire records to identify particle pollution originated from biomass burning
Velasco and Rastan (2015)	PM_{10}, $PM_{2.5}$	Singapore	Jun 2013	Likelihoods estimates of hourly particle concentrations experienced during the 2013 smoke-haze episode. At that time, no hourly air quality data was available	24-h moving average $PM_{2.5}$ concentrations reported by authorities and a set of numerical, stochastic and statistical models were used to calculate 1-h $PM_{2.5}$ likelihoods
Latif et al. (2014)	O_3, NO_2, NO, PM_{10}, total HC, CH_4	Jerantut, Malaysia	1997–2011	Information on regional air quality	Data from an official regional air quality monitoring station. Statistical analysis, including cluster analysis, sensitivity analysis, and principal component regression

(continued)

Table 9.1 (continued)

Reference	Pollutant species/parameters	Study area	Study period	Topic	Approach
Particle chemical and physical characterization					
Dominutti et al. (2024)	$PM_{2.5}$, EC, OC, major ions, organic acids, sugars, sugar alcohols, anhydrosugars, trace elements, HuLiS, PAH, and oxidative potential	Hanoi, Vietnam	Sep 2019–Dec 2020	Chemical characterization of particle pollution and its oxidative potential	Source apportionment analysis of $PM_{2.5}$ on 24-h samples collected at an urban site and measurements of oxidative potential
Lu et al. (2024)	$PM_{2.5}$ and reactive oxygen species attached to particles	Singapore	Oct 30, 2019–Mar 23, 2020	Characterization of particle-bound reactive oxygen species in oil-based cooking emissions	Application of a fluorescence probe technique to determine peroxidic compounds on $PM_{2.5}$ filter samples collected from real kitchen environments and heated cooking oil emissions in laboratory
Hou et al. (2023)	Particle number concentration, CO, CO_2, CH_4, SO_2	Singapore	May 2018–Apr 2019	Petroleum industry's impact on the abundance of ultrafine particles	Particle number size distributions were measured using a scanning mobility particle sizer
Lorenzo et al. (2023)	AOD	Manila, Philippines	Jan 2009–Oct 2018	Aerosol climatology	Clustering analysis of AERONET AOD data and MERRA-2 meteorology and particle chemical composition data
Makkonen et al. (2023)	$PM_{2.5}$, water-soluble ions, trace elements, OC, EC, sugar anhydrides	Hanoi, Vietnam	Aug 2019–Jul 2020	Chemical composition and potential sources of $PM_{2.5}$	Aerosol sampling at two locations. Gravimetric measurements combined with offline chemical analysis

(continued)

Table 9.1 (continued)

Reference	Pollutant species/parameters	Study area	Study period	Topic	Approach
Promsiri et al. (2023)	$PM_{2.5}$, OC, EC, water-soluble organic carbon (WSOC), water-soluble ions, trace elements, heavy metals, arsenic, PAHs	Hat Yai, Thailand	Jun 2019–Nov 2020	Transboundary haze impacts from peatland fires and local emission sources on particle pollution	24-h sampling and offline chemical analysis. Combination of air mass backward trajectories and chemical mass balance source apportionment to identify particle sources
Hilario et al. (2022)	Size-segregated particles, 16 water-soluble ions, 27 trace elements, BC, AOD	Manila, Philippines	Jul 2018–Oct 2019	Wet scavenging efficiency to reduce particle abundance during the rainy season	Comprehensive analysis of particle-rain relationships based on size-resolved aerosol composition (48-h samples), AOD, and meteorological parameters
Kusumaningtyas et al. (2022)	AOD, Ångström exponent, precipitable water vapor, and single scattering albedo	Fire prone areas: Palangka Raya, Pontianak, and Jambi; Urban/Industrial areas: Bandung, Indonesia	Fire prone areas: 2012–2020. Urban/Industrial areas: 2009–2018	Seasonal variation of aerosol optical and radiative properties	AERONET version 3 data was used. Multiple clustering techniques were used to study the aerosol type in relation to climatological season using relationships among various aerosol optical and radiative parameters
Siregar et al. (2022)	$PM_{2.5}$, $PM_{2.5-10}$, water-soluble ions, trace elements, BC	Riau, Indonesia	Apr 21–5 Jul, 2014	Particle chemical characterization during forest and peatland fires	Aerosol sampling at two locations. Gravimetric measurements and offline chemical analysis

(continued)

Table 9.1 (continued)

Reference	Pollutant species/parameters	Study area	Study period	Topic	Approach
Zhang et al. (2022)	$PM_{2.5}$, OC, EC, WSOC, water-soluble ions	Yangon and Mandalay, Myanmar	Yangon: Dec 4–21, 2016, & Apr 10,28, 2017. Mandalay: Dec 26, 2016–Jan 15, 2017, & Mar 20–Apr 6, 2017	Concentration and composition of $PM_{2.5}$ during dry periods	4-h samples on filters and offline chemical analysis
Fujii et al. (2021)	WSOC and carbon content of HULIS in TSP, OC, EC, and levoglucosan	Kuala Lumpur, Malaysia	Jun 23–Jul 8, 2014	Humic-like substances in particles in Malaysia influenced by peatland fires	Ambient sampling of TSP and analysis by excitation emission matrix fluorescence spectroscopy
Lorenzo et al. (2021)	$PM_{2.5}$, water-soluble ions, trace elements and aerosol backscatter cross section	Manila, Philippines	Dec 31, 2019	Fireworks impact on particle pollution	Size-segregated sampling and offline chemical analysis. A high-spectral-resolution lidar was used to measure vertical profiles of aerosol backscatter. Particle morphology and chemical composition were investigated using high-resolution scanning electron microscopy in conjunction with energy dispersive X-ray analysis
Kongpran et al. (2021)	$PM_{2.5}$, PAHs	Northern Thailand	Mar 2018 & Jul 2018	Particles bounding PAHs during haze and non-haze periods	Gravimetric measurements at 12 sampling sites and chemical analysis by high-performance liquid chromatography (HPLC)

(continued)

Table 9.1 (continued)

Reference	Pollutant species/ parameters	Study area	Study period	Topic	Approach
Adam et al. (2020)	$PM_{2.5}$, BC, brown carbon (BrC), WSOC, mass absorption coefficient, mass absorption efficiency, Ångström absorption exponent	Singapore	May 2017 and Mar 2018	Characteristics of light absorbing carbonaceous aerosols	A suite of aethalometers and micro-aethalometers were used for measuring BC. $PM_{2.5}$ samples were collected on filters for BrC determination by spectral analysis (200–800 nm wavelength range) using a UV–VIS spectrophotometer
Braun et al (2020)	Size-segregated particles, water-soluble organic acids, inorganic ions, and water-soluble elements	Manila, Philippines	Jul–Oct, 2018	Long-range aerosol transport and its effects on size-resolved aerosol composition	Various data sources, including models, remote sensing, and in situ measurements
Kasthuriarachchi et al. (2020)	OC, EC, organic aerosol components, aerosols light absorbing properties	Singapore	May 14–Jun 9, 2017	BrC light absorbing properties	Positive matrix factorization (PMF) analysis of aerosol mass spectrometer measurements in conjunction with aerosol light absorbance measurements
Phairuang et al. (2020)	PM_{10}, $PM_{10-2.5}$, $PM_{2.5-1}$, $PM_{1-0.5}$, $PM_{0.5-0.1}$, EC, OC, and four OC fractions (OC1, OC2, OC3, OC4)	Hat Yai, Thailand	Jan–Dec 2018	Carbonaceous aerosol seasonal variations, including size-fractionated particles	48-h sampling with a cascade sampler, followed by offline chemical analysis
Rivellini et al. (2020)	$PM_{2.5}$, nonrefractory aerosols including organic and inorganic aerosols, refractory BC, OC, EC, trace metals, CO, O_3, NO_2	Singapore	May 14–Jun 9, 2017	Carbonaceous aerosol chemical characterization, emission sources, and aging processes	Soot particle aerosol mass spectrometer measurements and PMF analysis

(continued)

Table 9.1 (continued)

Reference	Pollutant species/ parameters	Study area	Study period	Topic	Approach
Santoso et al. (2020)	$PM_{2.5}$, $PM_{10-2.5}$, BC, trace elements	16 large Indonesian cities	2010–2017	Urban particle pollution	Weekly 24-h samples and offline chemical analysis
Stahl et al. (2020)	Size-segregated particles, water-soluble organic acids, methanesulfonate (MSA), inorganic ions, water-soluble elements, BC	Manila, Philippines	Jul 2018–Oct 2019	Sources apportionment and properties of size-resolved particulate organic acids and MSA	Ambient sampling using a 10-stage impactor. Offline chemical analysis using ion chromatography and a triple quadrupole inductively coupled plasma mass spectrometer. PMF and correlation analysis were conducted to investigate sources of organic acids and MSA
Cruz et al. (2019)	11-stage particle size-resolved up to 0.05 µm, water-soluble ions, trace elements	Manila, Philippines	Jul–Oct, 2018	Size-resolved composition and particle morphology	Chemical composition and morphology of size-resolved ambient particles were analyzed by gravimetry, ion chromatography, triple quadrupole inductively coupled plasma mass spectrometry, black carbon spectroscopy, and microscopy. Source apportionment analysis by PMF
Tham et al. (2019)	$PM_{2.5}$, EC, OC, OC fractions (OC1–OC4), EC fractions (EC1-EC3), water-soluble ions, levoglucosan, malic acid, oxalate	Singapore	May 2012–Jun 2013 and Ju 2015–Dec 2015	Carbonaceous particle profile associated with peat biomass burnings	24-h aerosol sampling on filters, gravimetric measurements, and chemical analysis through different offline methods

(continued)

Table 9.1 (continued)

Reference	Pollutant species/parameters	Study area	Study period	Topic	Approach
Thepnuan et al. (2019)	$PM_{2.5}$, EC, OC, anhydro-sugars, water-soluble ions, carboxylates, sugar, sugar alcohols	Chiang Mai, Thailand	Feb 23–Apr 28, 2016	Molecular markers of biomass burning in $PM_{2.5}$ during dry season haze episodes	Aerosol sampling on filters, gravimetric measurements, and chemical analysis through different offline methods
Zong et al. (2019)	Particle number concentration and size distribution (5–1000 nm)	Singapore	Jul 31–Oct 8, 2017	Size spectra and source apportionment of fine particles	High time-resolution measurements of particle number concentration and size distribution using a fast-response differential mobility spectrometer
Alas et al. (2018)	Equivalent black carbon (eBC)	Manila, Philippines	Apr–Jun 2015	Spatial and diurnal variability of eBC	Fixed and mobile measurements of eBC were made using a multi-angle absorption photometer at fixed sites and a micro-aethalometer during mobile measurements
Budisulistiorini et al. (2018)	PM_1, $PM_{2.5}$, nonrefractory aerosols including organic and inorganic aerosols, EC, OC, water-soluble OC, and inorganic ions	Singapore	Oct 10–31, 2015	Real-time observations of non-refractory submicron particles during a smoke-haze episode	Aerosol mass spectrometer measurements
Pani et al. (2018)	$PM_{2.5}$, EC, OC, AOD, single scattering albedo, Ångström exponent, sky radiance	Chiang Mai, Thailand	Mar 8–Apr 7, 2014	Chemical composition of biomass-burning aerosols and their microphysical, optical, and radiative properties over an urban atmosphere during the dry season	Integration of ground-based measurements, satellite retrievals, and application of a radiative transfer model during the 7-SEAS/BASELInE study

(continued)

Table 9.1 (continued)

Reference	Pollutant species/ parameters	Study area	Study period	Topic	Approach
Shi et al. (2018)	$PM_{2.5}$	South and Southeast Asia	1999–2014	Trends and spatial patterns of satellite-retrieved $PM_{2.5}$ concentrations	Long-term series of satellite-retrieved ground-level $PM_{2.5}$ concentrations with high spatial resolution ($0.01° \times 0.01°$) and application of the GEOS-Chem chemical transport model based on AOD retrievals from the MODIS, MISR and SeaWiFS satellite instruments
Wiggins et al. (2018)	$PM_{2.5}$, total carbon content, $\Delta^{14}C$	Singapore	Sep 2014–Oct 2015	Radiocarbon (^{14}C) measurements of carbonaceous aerosol emitted from wildfires on the Indonesian islands of Sumatra and Borneo	Weekly and daily ambient particle samples on filters. Total carbon content and ^{14}C were analyzed using a carbon cycle accelerator mass spectrometer
Kecorius et al. (2017)	eBC and refractory particle number size distributions	Manila, Philippines	May 16–Jun 11, 2015	Physical properties of traffic-related carbonaceous particles in a busy street canyon	Particle size distribution and size segregated mixing state were determined using a mobility particle size spectrometer and a volatility tandem differential mobility analyzer, respectively
Sulong et al. (2017)	$PM_{2.5}$, trace elements, water-soluble ions	Kuala Lumpur, Malaysia	Jun 2015–Jan 2016	Particle concentration, composition, source apportionment and health risk during haze and non-haze episodes	Aerosol sampling on filters, gravimetric measurements, and chemical analysis by offline methods. PMF was used for source apportionment analysis. The numerical model NAME and Global Fire Assimilation System (GFAS) were used to investigate particles origin

(continued)

Table 9.1 (continued)

Reference	Pollutant species/ parameters	Study area	Study period	Topic	Approach
Khan et al. (2016)	Water-soluble ions, trace elements, rare earth elements, EC and OC in $PM_{2.5}$	Kuala Lumpur, Malaysia	Jun 24–Sep 14, 2014	Particle physicochemical properties during the dry season	Back trajectory analysis and source apportionment by PMF on 24-h aerosol samples analyzed offline
Wang et al. (2015)	AOD, single scattering albedo, Ångström exponent, sky radiance	Northern Southeast Asia	Mar 1–Apr 15, 2014	Aerosol optical properties and vertical distributions in a major biomass-burning emission region	Measurements from four AERONET sun-sky radiometers and one lidar
Betha et al. (2014)	$PM_{2.5}$, 4-stage sized segregated particles, and 16 trace elements (B, Ca, K, Fe, Al, Ni, Zn, Mg, Se, Cu, Cr, As, Mn, Pb, Co, and Cd)	Singapore	Jun 20–Jul 28, 2013, and 12 Sep 12–Oct 2, 2013	Characterization of the chemical fractionation of particulate-bound trace elements and associated health risk during a smoke-haze episode and a unaffected period	The different forms of trace metals were extracted from size-segregated particle samples using a four-step sequential extraction procedure. The metal extracts were analyzed with an inductive coupled plasma mass spectrometer. A multiple-path particle dosimetry model was used to estimate the deposition of inhaled trace element
Betha et al. (2013)	12 trace metals in $PM_{2.5}$ (Al, Cd, Co, Cr, Cu, Fe, Mn, Ni, Pb, Ti, V and Zn)	Kalimantan, Indonesia	Sep–Oct, 2009	Chemical speciation of trace metals emitted from peat fires for health risk assessment	Samples were collected in the immediate vicinity of peat fires. Offline analysis by inductively coupled plasma mass spectrometry (ICP-MS)

(continued)

Table 9.1 (continued)

Reference	Pollutant species/ parameters	Study area	Study period	Topic	Approach
Volatile organic compounds					
Hien et al. (2022)	33 nonmethane hydrocarbons (C_2–C_{12}), 6 oxygenated VOCs, monoterpenes, acetonitrile	Ho Chi Minh City and Hanoi, Vietnam	Ho Chi Minh City: Sep 27–Nov 15, 2018, and Mar 4–Apr 3, 2019 Hanoi: March 7–Apr 1, 2019	VOCs atmospheric composition and their role in ground-level ozone formation	Continuous sampling and analysis using online systems, including GC-FID, PTR-MS, and selected ion flow tube mass spectroscopy
Zulkifli et al. (2022)	29 VOC species	Kuala Lumpur, Malaysia	Aug 2017–Jul 2018	VOC atmospheric composition	1-h sampling and online analysis by GC-FID
Ly et al. (2020)	53 VOC species	Hanoi, Vietnam	Dec 2014–Jan 2015	Characterization of roadside VOCs	Roadside sampling and offline analysis by GC-FID
Aung et al. (2019)	21 VOC species, 15 carbonyls, NO_2, NH_3, O_3	Yangon, Myanmar	May 5–11, 2017	Air pollutants initial assessment	Diffusive air samplers and offline chemical determination
Hamid et al. (2019)	Benzene, toluene, ethylbenzene, xylene isomers (BTEX)	Kuala Lumpur, Penang, Bangi, Langkawi and Danum Valley, Malaysia	Selected days between Jul 31 and Dec 6, 2015	Ambient BTEX concentrations	Active sampling with sorbent tubes. Samples were analyzed using thermal desorption coupled with GC–MS
Phuc and Oanh (2018)	BTEX	Hanoi, Vietnam	Jan 13–Feb 9 and Oct 13–Nov 9, 2015	Ambient and roadside BTEX concentrations	1-h or 2-h samples collected using charcoal tubes. Offline analysis by GC-FID

(continued)

Table 9.1 (continued)

Reference	Pollutant species/parameters	Study area	Study period	Topic	Approach
Barletta et al. (2017)	63 C_2–C_{10} NMHCs, CH_4, CO	Singapore	Aug 16–Nov 1, 2012	Snap-shot of VOC atmospheric composition	85 samples in canisters from 9 to 19 h. Offline analysis by GC-FID, electron capture detector, and quadrupole mass spectrometer detector
Tunsaringkarn et al. (2014)	BTEX and 12 carbonyl species	Bangkok, Thailand	Selected days between Feb–Jun 2013	Ambient concentrations of BTEX and carbonyls	8-h samples collected using charcoal tubes and 2,4-dinitrophenylhydrazine cartridges, and offline analysis by GC-FID and HPLC for BTEX and carbonyls, respectively
Emissions characterization					
Dominutti et al. (2023)	30 VOC species, CO, CO_2	Hanoi and Ho Chi Minh City, Vietnam	Ho Chi Minh City: Sep–Oct 2018, and Hanoi: Mar 2019	Emission profiles and emission factors of VOCs from passenger cars, buses, trucks, motorcycles, 3-wheeled motorcycles, domestic waste burning, and charcoal burning in street food stalls	Direct source emission measurements by grabbing samples in canisters and chemical analysis using a dual-channel GC-FID. Emission factors were determined using the carbon balance method
Chen et al. (2022)	Particle number size distribution, chemical composition, hygroscopicity	Laboratory. Peatland samples from Riau, Indonesia	–	Chemical aging of fresh peatland burning particles under elevated relative humidity	Oxidation experiments on primary organic aerosols generated from peat and vegetation smoldering combustion using an aerosol mass reactor coupled with a tandem differential mobility analyzer and a time-of-flight aerosol chemical speciation monitor

(continued)

Table 9.1 (continued)

Reference	Pollutant species/parameters	Study area	Study period	Topic	Approach
Phuong et al. (2022)	TSP, PM_{10}, $PM_{2.5}$, pPAHs, 11 VOCs, CO_2, NO_2, SO_2	Mar and May of 2018 and 2019	Vietnam's Mekong Delta	Emission factors for rice straw open burning	Open burning experiments *in situ*. Different sampling techniques and analytical instruments were employed. The carbon balance method was used to calculate emission factors
Yokelson et al. (2022)	100 NMHCs, CO_2, CO, CH_4	Kalimantan and Sumatra, Indonesia	Sep–Nov 2019	Emission factors of trace gases from peat fires and comparison with emission factors from previous studies	Direct ground-based field measurements of trace gases by whole air sampling in canisters and chemical determination by GC-FID, electron capture detection, and quadrupole mass spectrometer detection. The carbon balance method was used to derive emission factors
Akbari et al. (2021)	$PM_{2.5}$, and $PM_{2.5}$-bound metals (K, Na, Mg, Cr, Zn)	Upper Northern Thailand	—	Emission factors for metals bound with $PM_{2.5}$ and ashes from biomass burning, including agricultural residue and forest leaf litter	Emission factors were obtained using a combustion chamber equipped with particle collectors. Chemical analysis was carried out using inductively coupled plasma-optical emission spectrophotometry
Pham et al. (2021)	CO, CO_2, SO_2, NO_2, $PM_{2.5}$, pPAHs	Hanoi, Vietnam	2016–2018 burning seasons	Emission factors for rice straw burning	Emission factors were determined from hood and field experiments simulating the burning of small rice straws piles commonly used by farmers

(continued)

Table 9.1 (continued)

Reference	Pollutant species/ parameters	Study area	Study period	Topic	Approach
Sofwan and Latif (2021)	CO, CO_2, NO_X	Kuala Lumpur, Malaysia	Nov–Dec 2019	Emission factors of on-road vehicular fleet	Tailpipe measurements using a commercial exhaust gas analyzer. Emission factors were calculated using the fuel-carbon approach
Chantara et al. (2019)	CO, NO, NO_2, SO_2, $PM_{2.5}$, water-soluble ions, anhydrosugars (levoglucosan mannosan and galactosan), sugars (glucose and mannose) and sugar alcohols (myo-inositol, erythritol, arabitol and mannitol)	Upper Northern Thailand	Nov-Dec 2015, and Jan-Mar 2016	Emission factors for open burning of agricultural residue (rice straw and maize) and forest leaf litter	An open-system combustion chamber was built to simulate open biomass burning under field conditions. Different sampling techniques and analytical instruments were employed. The carbon balance method was used to calculate emission factors
Madueño et al. (2019)	Particle number concentration, eBC	Manila, Philippines	May 18–Jun 10, 2015	Emission factors of on-road vehicular fleet	A suite of online instruments stationed along the roadside was used to measure particle number concentration, size distribution, and eBC. The emission factors were calculated using an inverse modeling approach
Nghiem et al. (2019)	CO, CO_2, total hydrocarbons, NO_X	Hanoi, Vietnam	–	Emission factors for buses	Laboratory emission measurements under controlled conditions based on the local driving cycle

(continued)

Table 9.1 (continued)

Reference	Pollutant species/parameters	Study area	Study period	Topic	Approach
Sothea and Oanh (2019)	PM, EC, OC, CO, CO_2, NOx, SO_2, 16 PAHs (in gas, and PM phases), 10 water-soluble ions, 30 trace elements	Phnom Penh, Cambodia	–	Emission factors and profiles for backup diesel generators used in the commercial sector	Isokinetic sampling in an experimental hood. Different sampling techniques and analytical instruments were used. Emission factors were calculated using the total volume of flue gas released and the mass concentration of pollutants
Jayarathne et al. (2018)	70 particulate-phase species in $PM_{2.5}$ including EC, OC, WSOC, water-soluble ions, metals, and organic species	Kalimantan, Indonesia	Oct–Nov 2015	Emission factors for aerosols from smoldering peat fires	Particle samples were collected on filters from smoldering peats. Offline chemical determination using various analytical instruments. The emission factors were calculated using CO emissions as a baseline
Sirithian et al. (2018)	36 VOC species including alkanes, alkenes, oxygenated VOCs, halogenated VOCs, aromatic VOCs, and carbon disulfide	Tak, Thailand	–	Emission factors of speciated VOCs from maize residue open burning	A combustion chamber was used. Samples were collected in Tedlar bags and then transferred into canisters for chemical analysis by GC–MS. Emission factors were determined on a dry weight of biomass basis

(continued)

Table 9.1 (continued)

Reference	Pollutant species/parameters	Study area	Study period	Topic	Approach
Smith et al. (2018)	CO_2, CO, CH_4, NH_3, acetic acid, hydrogen cyanide, methanol, ethylene, ethane, formaldehyde, formic acid, acetylene	Peninsular Malaysia	Aug 2015 and Jul 2016	Emission factors of trace gases from peat fires including information on the physical properties of peat fuel (e.g., bulk density and fuel moisture)	Direct ground-based field measurements of trace gases and aerosols by Fourier transform infrared spectroscopy. Emission factors were derived from emission ratios using the carbon mass balance method and CO emission factors as reference
Stockwell et al. (2016)	90 pollutant species, including CO_2, CO, CH_4, NMHCs up to C_{10}, 15 oxygenated organic compounds, NH_3, HCN, NO_x, OC, BC	Kalimantan, Indonesia	Oct–Nov 2015	Emission factors of trace gases and aerosols from peat fires	Direct ground-based field measurements of trace gases and aerosols by Fourier transform infrared spectroscopy, whole air sampling in canisters, photoacoustic extinctiometers, and analyses of particulate filters. Emission factors were derived from emission ratios using the carbon mass balance method
Oanh et al. (2015)	BTEX, SO_2, NO_2, aldehydes, semi-VOCs including PAHs, organochlorinated pesticides (OCPs), and polychlorinated biphenyls (PCBs)	Pathumthani, Thailand	Dec–Apr, 2003–2006	Emission factors and emission profiles of gaseous and semi-volatile organic compounds emitted from field burning of rice straw	The emissions were characterized through open burning experiments in paddy fields and simulated burning in a laboratory hood. Different sampling techniques and analytical instruments were used. The carbon balance method was used to calculate emission factors
Fujii et al. (2014)	$PM_{2.5}$, OC, EC, levoglucosan, mannosan	Riau, Indonesia	May 16–17 and Jun 13–17, 2012	Characterization of carbonaceous aerosols emitted from peat fires	Direct sampling of particles at fire hotspots and offline chemical determination

(continued)

Table 9.1 (continued)

Reference	Pollutant species/parameters	Study area	Study period	Topic	Approach
Personal exposure					
Collado et al. (2023)	$PM_{2.5}$	Manila Philippines	Nov 12–Dec 15, 2018	Detailed characterization of Jeepney (small public buses) drivers' exposure	Portable monitors were used to collect data along a busy road
Huy et al. (2022)	PM_{10}, $PM_{2.5}$, PM_1	Ho Chi Minh City, Vietnam	Aug–Oct 2020	Particle exposure while commuting by motorcycle (with and without a face mask), car, and bus	A battery-powered portable monitor was used to collect data along a predetermined route
Madueño et al. (2022)	eBC	Manila, Philippines	Nov 2019–Mar 2020	eBC particle deposition in the respiratory tract while commuting	Exposure measurements were taken on 40 volunteers while they were commuting by public transportation and walking, using a system that measures eBC within the exhaled airflow
Velasco et al. (2022)	$PM_{2.5}$, eBC, particle number concentration, active surface area (ASA), pPAHs	Ho Chi Minh City, Vietnam	Feb 24–Mar 2, 2016	Particle characterization at the street level, and effectiveness of wearing face masks against traffic particles	Particle characterization at the roadside in seven roads and one underground parking lot during high traffic periods using portable monitors. The ability to filter particles of six representative masks mounted on manikins was tested at the curb of two busy roads
Chaiklieng (2021)	BTEX	Khon Kaen, Thailand	Jun–Jul 2018	Characterization of workers' exposure at gas stations	47 gasoline stations were sampled using charcoal tubes. GC-FID was used for chemical determinations

(continued)

Table 9.1 (continued)

Reference	Pollutant species/parameters	Study area	Study period	Topic	Approach
Quang et al. (2021)	eBC	Hanoi, Vietnam	2 weeks of Sep (year no indicated)	Exposure to eBC while commuting	Simultaneous measurements were taken with a pair of potable micro-aethalometers during trips by motorcycle and bus, as well as by motorcycle and private car
Velasco and Segovia (2021)	$PM_{2.5}$, eBC, particle number concentration, ASA, pPAHs	Singapore	Mar 2018	Particle exposure on double-decker buses when traveling on the upper and lower deck	Particle measurements were taken simultaneously on both decks using portable monitors
Fandi et al. (2020)	BTEX	Kuala Lumpur, Malaysia	Sep 2017–Jan 2018	Characterization of traffic officers' exposure	Personal samplers equipped with Tenax GR sorbents. GC–MS chemical analysis combined with thermal desorption
Tan et al. (2017)	$PM_{2.5}$, eBC, particle number concentration, ASA, pPAHs, CO	Singapore	30 days between Apr 10–Jun 24, 2013	Particle exposure and inhaled dose during commuting	Measurements were taken using a set of portable monitors during door-to-door trips walking and using motorized transport modes (subway, bus, taxi) along a selected route in a commercial district. These measurements were compared to those taken at a site not affected by traffic emissions
Velasco and Tan (2016)	$PM_{2.5}$, eBC, particle number concentration, ASA, pPAHs, CO	Singapore	15 days between Jan 2011–Jul 2012	Particle exposure at bus stops	Measurements were conducted using portable monitors during the morning and evening rush hour at five representative bus stops
Velasco et al. (2013)	eBC, CO	Bangkok, Thailand	Selected days May–Jun 2012	Commuter exposure on the mass transport *khlong* boats	Measurements were taken at a busy pier and on boats during commutes using portable monitors

(continued)

Table 9.1 (continued)

Reference	Pollutant species/ parameters	Study area	Study period	Topic	Approach
Epidemiology					
Ho et al. (2022)	PM_{10}, $PM_{2.5}$, NO_2, CO, SO_2, O_3	Singapore	Jul 2010–Dec 2018	Associations between out-of-hospital cardiac arrest (OHCA) and air pollution	Time-series regression analysis with multivariable fractional polynomial modeling and OHCA data from the Pan-Asian Resuscitation Outcomes Study, and pollutant concentrations from the local environmental agency
Tan et al. (2022)	PM_{10}, $PM_{2.5}$, O_3, NO_2, SO_2, CO	Singapore	2009–2018	Effect of atrial fibrillation on the association of acute ischemic strokes (AIS) with air pollution	Multivariable logistic regression adjusted for time-varying meteorological effects using daily cases of AIS from all clinical institutions and pollutant concentrations reported by the local environmental agency
Chujit et al. (2020)	PM_{10}, $PM_{2.5}$, O_3, NO_2, SO_2, CO	Lampang, Thailand	Nov 2015–Oct 2016	Relationship between air pollutants and peak expiratory flow rates (PEFRs) in adults with asthma	Statistical models, a panel study of 33 asthmatic adults, and air quality data from a local air quality monitoring station
Luong et al. (2020)	$PM_{2.5}$	Ho Chi Minh City, Vietnam	Feb 2016–Dec 2017	Risk of hospital admission for acute lower respiratory infection in children due to particle pollution	Time-series regression analysis using data from hospitals and four air quality monitoring stations

(continued)

Table 9.1 (continued)

Reference	Pollutant species/parameters	Study area	Study period	Topic	Approach
Yu et al. (2020)	$PM_{2.5}$	Yogyakarta, Indonesia	Mar–Apr, 2016	Long-term particle exposure and fasting plasma glucose (FPG) in adolescents	$PM_{2.5}$ data from satellite reanalysis products and blood samples from 469 participants after overnight fasting. A generalized linear regression model was used to determine the relationship between $PM_{2.5}$ and FPG
Ho et al. (2019)	Singapore's Pollutant Standard Index	Singapore	2010–2015	Short-term air pollution exposure and the incidence of acute myocardial infarction in the context of smoke-haze episodes	Time-stratified case-crossover study of all acute myocardial infarction reported in Singapore and the local Pollutant Standard Index
Nhung et al. (2019)	PM_{10}, $PM_{2.5}$, PM_1, SO_2, NO, NO_2, CO, O_3	Hanoi, Vietnam	2007–2016	Association between ambient air pollution and length of hospital stay for children with lower-respiratory infection	Odds ratio of being discharged for an interquartile range increment of pollutant concentrations during one to four days prior to admission date adjusting for diverse factors and using data from Vietnam National Children's Hospital
Phosri et al. (2019)	PM_{10}, O_3, NO_2, SO_2, CO	Bangkok, Thailand	2006–2014	Short-term effects of ambient air pollutants on the risk of daily hospital admissions for cardiovascular and respiratory diseases	Time-series regression analysis using city-daily hospital admissions data from Thailand's National Health Security Office of Thailand and pollutants concentration data from the local air quality monitoring network

(continued)

Table 9.1 (continued)

Reference	Pollutant species/ parameters	Study area	Study period	Topic	Approach
Luong et al. (2017)	PM_{10}, $PM_{2.5}$, PM_1	Hanoi, Vietnam	Sep 2010–Sep 2011	Short-term effects of particle pollution on pediatric hospital admissions for respiratory problems	Time-stratified case-crossover analysis using data from one hospital and one air quality monitoring station
Pongpiachan et al. (2015)	pPAHs	Northern Thailand	Dec 2012 and Mar 2013	Potential cancer risk resulting from biomass burning pPAHs	Aerosol sampling, gravimetric measurements, and chemical analysis by HPLC. Cancer risk was estimated using algorithms available in the literature
Phung et al. (2016)	NO_2, PM_{10}, $PM_{2.5}$, SO_2, O_3	Ho Chi Minh City, Vietnam	Feb 2004–Dec 2007	Short-term effects of air pollutants on respiratory and cardiovascular hospitalizations	Time-series regression analysis using data from hospitals and one air quality monitoring station
Guo et al. (2014)	PM_{10}, SO_2, O_3	18 provinces across Thailand	1999–2008	Effects of exposure to air pollution on mortality risks	Bayesian statistical inference using a case-crossover design. Mortality and air quality data were obtained from government institutions
Mahiyuddin et al. (2013)	PM_{10}, SO_2, CO, NO_2, O_3	Kuala Lumpur, Malaysia	2000–2006	Short-term effects of daily air pollution on mortality	A time-series analysis based on a generalized additive model. Data on mortality and air quality were obtained from government agencies

(continued)

Table 9.1 (continued)

Health and economic costs associated with air pollution

Reference	Pollutant species/parameters	Study area	Study period	Topic	Approach
Bui and Nguyen (2023)	$PM_{2.5}$	Ho Chi Minh City, Vietnam	2018	Mortality, morbidity, and economic cost associated with particle pollution	Modeling of air quality and health impacts
Ho (2017)	PM_{10}	Ho Chi Minh City, Vietnam	Jul 2012	Mortality and economic costs associated with particle pollution-related health problems	Air quality modeling (FVM-TAPOM) and the USA-EPA BenMAP-CE open-software were used to estimate health and economic costs
Pinichka et al. (2017)	PM_{10}, $PM_{2.5}$, NO_2	Thailand	2009	Thailand's spatial pattern of mortality burden attributable to ambient air pollution	Comparative risk assessment integrated into a geographical information system. Relationships between relative risk and pollutant concentrations were obtained from epidemiological literature, while pollutants concentration from Thailand's air quality monitoring network
Yorifuji et al. (2015)	PM_{10}, $PM_{2.5}$	27 cities of Asia, including Caloocan, Manila, Davao, Hanoi, Ho Chi Minh City, Quezon City, and Singapore	2009	Annual health impact of particle pollution	Ambient concentrations from air quality monitoring stations and a health risk assessment based on pollutant relative risks available in the literature

(continued)

Table 9.1 (continued)

Reference	Pollutant species/ parameters	Study area	Study period	Topic	Approach
Othman et al. (2014)	PM_{10}, SO_2, CO, NO_2, O_3	Kuala Lumpur, Malaysia	2005–2006, and 2008–2009	Economic value of transboundary smoke-haze pollution's health impacts	Ambient concentrations were obtained from air quality monitoring stations, and inpatient data was obtained from hospitals. The cost illness approach and dose response functions based on panel data regression were used to calculate impact of smoke-haze on health and the economy
Toxicology					
Maciaszek et al. (2023)	PM_{10}	Chiang Mai, Thailand	2020	Pulmonary toxicity of PM_{10} emitted during biomass-burning smoke-haze episodes	In vitro assessment of particles toxicity using macrophage responses and analysis of haze and non-haze particle samples' physicochemical properties
Othman et al. (2022)	PM_{10}, $PM_{2.5}$, PM_1, CO_2	Kuala Lumpur, Sarawak, center of Peninsular Malaysia, and Chiang Mai, Thailand	2019–2020	Particle deposition fraction in the respiratory tract during biomass-burning episodes	Continuous particle measurements were taken using a special device built with low-cost sensors. The particle deposition fraction in the human respiratory tract was estimated using Multiple-path Particle Dosimetry software

(continued)

Table 9.1 (continued)

Reference	Pollutant species/ parameters	Study area	Study period	Topic	Approach
George et al. (2020)	Size fractionated PM and BC	Singapore	Jan–Dec 2014	Physicochemical, cytotoxicity and inflammatory potential of particle pollution during haze and non-haze episodes triggered by biomass burning	Continuous real-time ambulatory monitoring of BC by school children. Outdoor particle collection. Particles were characterized for size, morphology, elemental composition, trace metal content, ability to generate reactive oxygen species, cytotoxicity and inflammatory potential
Pavagadhi et al. (2013)	$PM_{2.5}$	Singapore	Oct 2010	Physicochemical and toxicological characteristics of urban aerosols affected by biomass-burning smoke-haze	10-day sampling period. Particles were analyzed for PAHs and trace metals. A human epithelial lung cell line was used to evaluate particles toxicity. Glutathione and caspase-3/7 levels were measured to assess oxidative stress and apoptotic death, respectively

Bangkok, Chiang Mai, Kuala Lumpur, and Singapore. In these cities, research has been conducted to a greater or lesser extent on the temporal variability of key pollutants, the impact of biomass burning during smoke-haze episodes, the chemical composition of the particles, as well as the impact of air pollution on public health and associated economic costs. Cambaliza et al. (2023) provide a summary of the most important findings from recent research studies conducted in Manila, Kuala Lumpur, and Singapore.

- Over the past few years, significant advancements in the chemical and physical characterization of particles have been made in these cities. We have information on the content of organic and inorganic carbon, black carbon, water-soluble ions, trace elements, and bounded PAHs that are present in particles (e.g., Braun et al. 2020; Cruz et al. 2019; Budisulistiorini et al. 2018; Khan et al. 2016). We even have size-resolved composition and morphology data of the particles for some of these cities (Lorenzo et al. 2021; Zong et al. 2019; Kecorius et al. 2017). A few studies have also analyzed the carbon content of HULIS (Fujii et al. 2021), the fractions of organic carbon (OC1-OC4) and elemental carbon (EC1-EC3) (Tham et al. 2019; Thepnuan et al. 2019), as well as conducted radiocarbon (^{14}C) measurements (Wiggins et al. 2018), which, when combined with measurements of molecular markers of biomass burning such as levoglucosan, anhydro-sugars, and carboxylates, among others chemical species, enable source apportionment analysis (e.g., Makkonen et al. 2023; Stahl et al. 2020; Thepnuan et al. 2019).

- Most of the particle studies have been conducted through the collection of particles on filters and subsequent offline chemical analysis. Only a few short (i.e., weeks) studies in Singapore have used online analytical instrumentation, such as aerosol mass spectrometers, to determine the non-refractory main species (nitrate, sulfate, ammonium, chloride, and organic compounds) in PM_1 and $PM_{2.5}$ (Kasthuriarachchi et al. 2020; Rivellini et al. 2020; Budisulistiorini et al. 2018). Online measurements have the advantage of avoiding losses of semi-volatile species, which occur during particle sampling on filters and chemical analysis. Nowadays, it is common for research institutions and even air quality monitoring networks to use commercial instruments, such as the Aerosol Chemical Speciation Monitor (ACSM, Aerodyne Research Inc., Billerica, MA, USA) and the Monitor for AeRosols and GAses (MARGA, Metrohm Applikon B. V., Netherlands) to measure online particle chemical composition..

- Studies in Singapore, Manila, and Chiang Mai in northern Thailand have evaluated the optical and radiative properties of the particles to determine their ability to absorb and scatter light, as well as to investigate the temporal variations of brown carbon (e.g., Lorenzo et al. 2023; Adam et al. 2020; Pani et al. 2018). These studies have also been conducted in rural areas of Indonesia affected by peatland fires (e.g., Kusumaningtyas et al. 2022).

- Despite the relevance of volatile organic compounds (VOCs) in the formation of secondary pollutants such as O_3 and secondary organic aerosols, little attention has been paid to measuring them. Until now, only the findings of three comprehensive studies have been published. One study measured ambient concentrations in Hanoi and Ho Chi Minh City (Hien et al. 2022), another in Kuala Lumpur (Zulkifli et al.

2022), and a third measured roadside concentrations in Hanoi (Ly et al. 2020). These studies measured between 29 and 53 species. The first two studies used online gas chromatography by flame ionization (GC-FID) to conduct continuous measurements. Proton-transfer reaction mass spectrometry (PTR-MS) was also used in the study in both Vietnamese cities. The study in Hanoi's streets relied on collecting samples and conducting offline analysis. Other VOCs studies in the region have either been snapshots of VOCs speciation (Barletta et al. 2017) or have been limited to measuring so called BTEX (benzene, toluene, ethylbenzene and xylene isomers) (e.g., Hamid et al. 2019; Tunsaringkarn et al. 2014). In the case of Singapore, Cambaliza et al. (2023) brought up a comprehensive study in which over 200 species were analyzed between 2018 and 2020. The overall results are provided, and their potential to form O_3 and secondary organic aerosols is even discussed, but no details of the measurements and specific results were included.

- Direct field measurements and combustion chamber experiments have been used to determine VOC emission factors and profiles from biomass burning, both for agricultural residues (e.g., Phuong et al. 2022; Sirithian et al. 2018; Oanh et al. 2015) and peatland fires (e.g., Yokelson et al. 2022; Smith et al. 2018; Stockwell et al. 2016). Only one study has determined emission factors and profiles from passenger cars, buses, trucks, motorcycles, and three-wheeled motorcycles (Dominutti et al. 2023). This study was based on online observations of VOCs near busy roadways and direct source emission measurements in Hanoi and Ho Chi Minh City streets. Emission factors for burning domestic waste and charcoal used as fuel by street food stalls were also determined.

- Very few emission factors for pollutant species from sources other than biomass burning have been locally determined. Regarding vehicle emissions, Madueño et al. (2019) determined the emission factors of equivalent black carbon (eBC) and particle number for Manila's fleet as a whole; Nghiem et al. (2019) determined the emission factors of CO, CO_2, NO_x, and total hydrocarbons for buses in Hanoi; and Sofwan and Latif (2021) determined the emission factors of CO, CO_2, and NO_x for on-road vehicles in Kuala Lumpur. For other emission sources, we only found the study by Sothea and Oanh (2019), who determined emission factors of pollutant gases and chemically speciated particles for backup diesel generators commonly used in Cambodia.

- Another pollutant that has received insufficient attention from the academic community is O_3. As part of their air quality programs, seven countries monitor ambient O_3 concentrations, mostly in urban areas. The collected data has been used to assess the diurnal and seasonal variations and spatial distribution of O_3 (e.g., Sukkhum et al. 2022; Nghiem et al. 2021; Halim et al. 2020), as well as changes in its diurnal pattern during periods of smoke-haze triggered by regional biomass burning (Awang et al. 2018). All of these studies have used statistical analyses, some more complex than others, but only Uttamang et al. (2018) has attempted to elucidate the mechanisms underlying O_3 formation. They estimated O_3 photolytic rates assuming a photo-stationary state using O_3, NO_2 and NO data from Bangkok's local air quality monitoring network to determine the contribution of O_3 transported regionally to O_3 formed locally.

- Ozone formation is a complex process that is driven by the local abundance of VOCs and NO_x in the presence of solar radiation. The hydroxyl radical (OH) reacts with VOCs to initiate the fast chemistry that transforms the NO_x and VOCs emitted into the atmosphere by a miyrd of emission sources. Following reactions generate the hydroperoxyl radical (HO_2), which reacts with NO to reform OH and produce NO_2, resulting in the formation of O_3. Low volatility VOCs are also generated, resulting in the formation of secondary organic aerosols. Together, OH and HO_2, known as HO_x, form a rapid reaction cycle that drives the O_3 chemistry. Here the importance of counting with locally derived information on OH, HO_2 and OH reactivity to understand O_3 production and design effective NO_x and VOCs emission control measures (Yang et al. 2016; Brune et al. 2016; Stone et al. 2012). Only Edwards et al. (2013) have investigated OH reactivity in Southeast Asia as part of a project focused on assessing photochemical processes in Borneo's tropical forests, which is mentioned below.

- Similarly, the processes of particle growth, formation, and destruction in the atmosphere have received little attention in Southeast Asia. Indeed, many of the chemical-transport models introduced in Chapter 7 and listed in Table 7.1 include modules for simulating heterogeneous chemistry (i.e., chemical processes between aerosols and gases). However, the lack of information on precursor species and HO_x radicals, as well as the chemical speciation of aerosols, has prevented evaluating the accuracy of the chemical mechanisms used in such modules. The conversion of atmospheric gases into particles, or vice versa, under specific thermodynamic conditions, results in the formation of new particles, as well as the growth and oxidation of existing particles. The process can take various paths depending on the relative abundance of precursors gases such as SO_2, NO_x, NH_3, and VOCs, atmospheric conditions (temperature, humidity, etc.), abundance of HO_x radicals, chemical and physical properties of already existing aerosols, and solar radiation, among other factors (Seinfeld and Pandis 2016). As a result, as in the case of O_3, it is critical to have information on HO_x radicals and OH reactivity, as well as the multiple precursor species involved in aerosol chemistry, in order to evaluate the accuracy of current chemical mechanisms and propose mechanisms that are appropriate for Southeast Asia's atmospheric conditions. Only one study has thoroughly investigated photochemical processes in the region through field measurements and modeling of atmospheric composition and chemistry in Borneo's tropical rainforest (Oxidant and particle photochemical processes above a South-East Asian tropical rainforest, OP3 Project, Hewitt et al. 2010).

- The impact of land cover changes experienced throughout the region in recent decades on atmospheric chemistry, beyond the impact caused by biomass burning, is a subject that needs further research. The increasing demand for palm oil, which is extracted from oil palm trees, has resulted in widespread expansion of this agro-industrial tree crop at the expense of large-scale deforestation, particularly in Indonesia and Malaysia. Oil palm plantations emit far more biogenic isoprene than local natural forests (Langford et al. 2010; Misztal et al. 2011), and this, combined with increased NO_x emissions from increased fertilizer use, increases the formation of O_3 and secondary organic aerosols, with significant implications for air quality and climate forcing (e.g., Harper and Unger 2018; Silva et al. 2016; Warwick et al. 2013; Hewitt et al. 2009). These perturbations in the emission of biogenic volatile organic compounds (BVOCs) are relevant at the local, regional, and global scale. On the local scale, cities such as Kuala Lumpur and Singapore may experience more days with high O_3 concentrations (Silva et al. 2016). Higher levels of O_3 and secondary organic aerosols on a regional scale may be harmful to natural ecosystems (Hewitt et al. 2009). And on a global scale, studies suggest that increased BVOCs may have a significant impact on the abundance of both pollutant species in the upper troposphere due to the strong vertical mixing prevalent in the tropics, gaining relevance for global radiative forcing (Harper and Unger 2018; Warwick et al. 2013). Most studies have focused on the last two aspects and less on the first.

- A number of studies have been conducted in the Philippines, Singapore, Thailand, and Vietnam on personal exposure to airborne pollutants at street level. These studies are important because they assess the amount of pollutants that people are exposed to during their daily outdoor activities. Remember that for regulatory and warning purposes, air quality monitoring stations measure pollutant concentrations at ambient level (i.e., above the urban canopy). Studies in Thailand and Malaysia have focused on measuring exposure to BTEX (e.g., Chaiklieng 2021; Fandi et al. 2020), whereas studies in the other three countries have focused on measuring exposure to particle pollution, particularly during commuting trips on different transport modes (e.g., Collado et al. 2023; Madueño et al. 2022; Quang et al. 2021; Velasco et al. 2013). All of these studies have measured $PM_{2.5}$ or equivalent black carbon (eBC), only the studies conducted in Singapore have expanded the measurements to include other particles metrics like number concentration (as a proxy of ultrafine particles, UFP), particles bound PAHs (pPAHs), active surface area (ASA), and mean particle size (Velasco and Segovia 2024, 2021; Tan et al. 2017; Velasco and Tan 2016). Two studies in Vietnam have looked at the effectiveness of wearing face masks on the street to reduce particle exposure (Huy et al. 2022; Velasco et al. 2022).

- To date, a handful of epidemiological studies conducted in Malaysia, Singapore, Thailand, and Vietnam have conclusively demonstrated the harmful health effects of breathing polluted air. Some of these studies have focused on long-term effects (e.g., Ho et al. 2022; Luong et al. 2017; Guo et al. 2014), while others have focused on the short-term effects (e.g., Ho et al. 2019; Phosri et al. 2019; Phung et al. 2016). They have all used data collected in hospitals and local air quality monitoring stations. They have consistently found associations between cardiovascular and respiratory illnesses, hospital visits and admissions, and mortality. It is worth noting a study conducted with teenagers in Yogyakarta, the only recent study found for Indonesia, in which long-term exposure to particle pollution and fasting plasma glucose as a proxy of diabetes risk were investigated (Yu et al. 2020).

- In recent years, the economic cost and mortality rate associated with air pollution have only been assessed at the city scale in Hanoi, Ho Chi Minh City, Kuala

Lumpur, Singapore, and four cities in the Philippines (Bui and Nguyen 2023; Ho 2017; Yorifuji et al. 2015; Othman et al. 2014). At country scale, beyond the estimates of the State of Global Air 2024 and the Global Burden of Diseases 2019 on premature deaths (HEI 2024; IHME 2020), and the estimates of the monetary cost of mortality and morbidity caused by air pollution (World Bank 2022), only one study in Thailand has done so as well. Pinichka et al. (2017) used a geographical system and data from Thailand's air quality monitoring network to conduct a comparative risk assessment to determine the spatial pattern of mortality attributable to air pollution across the entire country.

- Only three toxicology studies based on *in-vitro* tests on particulate toxicity were found in the literature. Two were made with particles collected in Singapore (George et al. 2020; Pavagadhi et al. 2013), and the third with particles collected in Chiang Mai (Maciaszek et al. 2023). These studies evaluated the physicochemical and toxicological properties of particles affected and not affected by biomass burning smoke-haze events. The cytotoxicity and inflammatory potential of the particles, as well as their responses to macrophages, and potential for oxidative stress, were investigated. A fourth study used a model and ambient particle concentrations measured in Chiang Mai and three Malaysian sites to assess the fraction of particle deposition in the respiratory tract (Othman et al. 2022).

- Thailand is most likely the country with the most air quality studies outside of its capital city. This is especially true for the upper northern region, which is seasonally affected by episodes of smoke haze caused by burning agricultural residues. In Malaysia, research has been conducted throughout the country, particularly in the peninsular region; however, the majority of studies have concentrated on Kuala Lumpur. Outside of Hanoi and Ho Chi Minh City, there is very little research work in Vietnam, as there is in the Philippines outside of Manila.

- Much of the research done so far in Indonesia has responded to international efforts to understand the chemical and physical processes in the atmosphere triggered by emissions from peatland fires in Sumatra and Kalimantan, which due to their magnitude, in addition to be a local problem, they are also a regional and global concern. Efforts to investigate any issue related to air quality on the rest of the archipelago's islands are incipient or null. Even Jakarta, home to 11 million people and one of the cities with the worst air quality in the region, has only one relevant scientific article related to air pollutant emissions (Lestari et al. 2022).

- Myanmar, Cambodia, Laos, and Timor-Leste are in even worse situation. Only three recent studies for Myanmar (Zhang et al. 2022; Huy et al. 2020; Aung et al. 2019) and one for Cambodia (Sothea and Oanh 2019) were found, with no studies for the other two countries. These countries lack the technical, human, and financial resources necessary to conduct research studies to better understand their air quality problems. Unfortunately, the international community has remained unconcerned about the threat that air pollution poses to the inhabitants of these countries.

- The amount of research done in Brunei Darussalam is not proportionate to its economic capacity. It is a small country, but it is the second wealthiest country in the region, and it suffers from the noxious smoke-haze caused by smoldering peat fires in Kalimantan almost every year. There are only three studies on air quality, all of which were conducted by the same group of researchers (Dotse et al. 2016a,b, 2018).

- Singapore has the technical and financial resources to conduct high-level scientific research. We can cite a handful of good studies in the field, but perhaps not as many as one might expect given its strong profile in advanced science and technological innovation. There are three main factors that have stymied further scientific advancement. To begin with, authorities consider air quality to be a sensitive topic, and as a result, it is not investigated as thoroughly as other environmental topics are. Second, there is a lack of collaborative work within local academia. And third, most of the researchers are not Singaporeans, and they generally see Singapore as a stopover in their academic journey, so their genuine concern for local issues is limited.

References

Adam, M.G., Chiang, A.W.J., Balasubramanian, R.: Insights into characteristics of light absorbing carbonaceous aerosols over an urban location in Southeast Asia. Environ. Pollut. **257**, 113425 (2020). https://doi.org/10.1016/j.envpol.2019.113425

Adam, M.G., Tran, P.T., Bolan, N., Balasubramanian, R.: Biomass burning-derived airborne particulate matter in Southeast Asia: a critical review. J. Hazard. Mater. **407**, 124760 (2021). https://doi.org/10.1016/j.jhazmat.2020.124760

Akbari, M.Z., Thepnuan, D., Wiriya, W., Janta, R., Punsompong, P., Hemwan, P., Charoenpanyanet, A., Chantara, S.: Emission factors of metals bound with $PM_{2.5}$ and ashes from biomass burning simulated in an open-system combustion chamber for estimation of open burning emissions. Atmos. Pollut. Res. **12**(3), 13–24 (2021). https://doi.org/10.1016/j.apr.2021.01.012

Alas, H.D., Müller, T., Birmili, W., Kecorius, S., Cambaliza, M.O., Simpas, J.B.B., Cayetano, M., Weinhold, K., Vallar, E., Galvez, M.C., Wiedensohler, A.: Spatial characterization of black carbon mass concentration in the atmosphere of a Southeast Asian megacity: an air quality case study for Metro Manila, Philippines. Aerosol Air Quality Res. **18**(9), 2301–2317 (2018). https://doi.org/10.4209/aaqr.2017.08.0281

Aung, W.Y., Noguchi, M., Yi, E.E.P.N., Thant, Z., Uchiyama, S., Win-Shwe, T.T., Kunugita, N., Mar, O: Preliminary assessment of outdoor and indoor air quality in Yangon city, Myanmar. Atmos. Pollut. Res. **10**(3), 722–730 (2019). https://doi.org/10.1016/j.apr.2018.11.011

Awang, N.R., Ramli, N.A., Shith, S., Zainordin, N.S., Manogaran, H.: Transformational characteristics of ground-level ozone during high particulate events in urban area of Malaysia. Air Quality Atmos. Health **11**, 715–727 (2018). https://doi.org/10.1007/s11869-018-0578-0

Barletta, B., Simpson, I.J., Blake, N.J., Meinardi, S., Emmons, L.K., Aburizaiza, O.S., Siddique, A., Zeb, J., Yu, L.E., Khwaja, H.A., Farrukh, M.A.: Characterization of carbon monoxide, methane and nonmethane hydrocarbons in emerging cities of Saudi Arabia and Pakistan and in Singapore. J. Atmos. Chem. **74**, 87–113 (2017). https://doi.org/10.1007/s10874-016-9343-7

Betha, R., Behera, S.N., Balasubramanian, R.: 2013 Southeast Asian smoke haze: fractionation of particulate-bound elements and associated health risk. Environ. Sci. Technol. **48**(8), 4327–4335 (2014). https://doi.org/10.1021/es405533d

Betha, R., Pradani, M., Lestari, P., Joshi, U.M., Reid, J.S., Balasubramanian, R.: Chemical speciation of trace metals emitted from Indonesian peat fires for health risk assessment. Atmos. Res. **122**, 571–578 (2013). https://doi.org/10.1016/j.atmosres.2012.05.024

Braun, R.A., Aghdam, M.A., Bañaga, P.A., Betito, G., Cambaliza, M.O., Cruz, M.T., Lorenzo, G.R., MacDonald, A.B., Simpas, J.B., Stahl, C., Sorooshian, A.: Long-range aerosol transport and impacts on size-resolved aerosol composition in Metro Manila, Philippines. Atmos. Chem. Phys. **20**, 2387–2405 (2020). https://doi.org/10.5194/acp-20-2387-2020

Brune, W.H., Baier, B.C., Thomas, J., Ren, X., Cohen, R.C., Pusede, S.E., Browne, E.C., Goldstein, A.H., Gentner, D.R., Keutsch, F.N., Thornton, J.A., Harrold, S., Lopez-Hilfiker, F.D., Wennberg, P.O.: Ozone production chemistry in the presence of urban plumes. Faraday Discuss. **189**, 169–189 (2016). https://doi.org/10.1039/C5FD00204D

Budisulistiorini, S.H., Riva, M., Williams, M., Miyakawa, T., Chen, J., Itoh, M., Surratt, J.D., Kuwata, M.: Dominant contribution of oxygenated organic aerosol to haze particles from real-time observation in Singapore during an Indonesian wildfire event in 2015. Atmos. Chem. Phys. **18**, 16481–16498 (2018). https://doi.org/10.5194/acp-18-16481-2018

Bui, L.T., Nguyen, P.H.: Evaluation of the annual economic costs associated with PM$_{2.5}$-based health damage-a case study in Ho Chi Minh City, Vietnam. Air Qual. Atmos. Health **16**, 415–435 (2023). https://doi.org/10.1007/s11869-022-01282-0

Cambaliza, M.O.L., NUS AQ Lab., Latif, M.T., Lestari, P.: Regional and urban air quality in Southeast Asia: maritime Continent. In: Akimoto, H., Tanimoto, H. (eds.) Handbook of Air Quality and Climate Change. Springer, Singapore (2023). https://doi.org/10.1007/978-981-15-2527-8_68-1

Chaiklieng, S.: Risk assessment of workers' exposure to BTEX and hazardous area classification at gasoline stations. PLoS ONE **16**(4), e0249913 (2021). https://doi.org/10.1371/journal.pone.0249913

Chantara, S., Thepnuan, D., Wiriya, W., Prawan, S., Tsai, Y.I.: Emissions of pollutant gases, fine particulate matters and their significant tracers from biomass burning in an open-system combustion chamber. Chemosphere **224**, 407–416 (2019). https://doi.org/10.1016/j.chemosphere.2019.02.153

Chen, J., Budisulistiorini, S.H., Itoh, M., Kuwata, M.: Roles of relative humidity and particle size on chemical aging of tropical peatland burning particles: Potential influence of phase state and implications for hygroscopic property. J. Geophys. Res. Atmos. **127**(14), e2022JD036871 (2022). https://doi.org/10.1029/2022JD036871

Chujit, W., Wiwatanadate, P., Deesomchok, A., Sopajaree, K., Eldeirawi, K., Tsai, Y.I.: Air pollution levels related to peak expiratory flow rates among adult asthmatics in Lampang, Thailand. Aerosol Air Qual. Res. **20**, 1398–1410 (2020). https://doi.org/10.4209/aaqr.2020.03.0092

Collado, J.T., Abalos, J.G., de los Reyes, I., Cruz, M.T., Leung, G.F., Abenojar, K., Manalo, C.R., Go, B., Chan, C.L., Gonzales, C.K.G., Simpas, J.B.B., Porio, E.E., Wong, J.Q., Lung, S.C.C., Cambaliza, M.O.L.: Spatiotemporal assessment of PM$_{2.5}$ exposure of a high-risk occupational group in a Southeast Asian Megacity. Aerosol Air Qual. Res. **23**(1), 220134 (2023). https://doi.org/10.4209/aaqr.220134

Cruz, M.T., Bañaga, P.A., Betito, G., Braun, R.A., Stahl, C., Aghdam, M.A., Cambaliza, M.O., Dadashazar, H., Hilario, M.R., Lorenzo, G.R., Ma, L., MacDonald, A.B., Pabroa, P.C., Yee, J.R., Simpas, J.B., Sorooshian, A.: Size-resolved composition and morphology of particulate matter during the southwest monsoon in Metro Manila, Philippines. Atmos. Chem. Phys. **19**(16), 10675–10696 (2019). https://doi.org/10.5194/acp-19-10675-2019

Dominutti, P.A., Hopkins, J.R., Shaw, M., Mills, G.P., Le, H.A., Huy, D.H., Forster, G.L., Keita, S., Hien, T.T., Oram, D.E.: Evaluating major anthropogenic VOC emission sources in densely populated Vietnamese cities. Environ. Pollut. **318**, 120927 (2023). https://doi.org/10.1016/j.envpol.2022.120927

Dominutti, P.A., Mari, X., Jaffrezo, J.L., Ngoc, T.V.D., Chifflet, S., Guigue, C., Guyomarc'h, L., Vu, C.T., Darfeuil, S., Ginot, P., Elazzouzi, R., Mhadhbi, T., Voiron, C., Uzu, G.: Disentangling fine particles (PM$_{2.5}$) composition in Hanoi, Vietnam: Emission sources and oxidative potential. Sci. Total Environ. **923**, 171466 (2024). https://doi.org/10.1016/j.scitotenv.2024.171466

Dotse, S.Q., Dagar, L., Petra, M.I., De Silva, L.C.: Evaluation of national emissions inventories of anthropogenic air pollutants for Brunei Darussalam. Atmos. Environ. **133**, 81–92 (2016a). https://doi.org/10.1016/j.atmosenv.2016.03.024

Dotse, S.-Q., Dagar, L., Petra, M.I., De Silva, L.C.: Influence of Southeast Asian haze episodes on high PM_{10} concentrations across Brunei Darussalam. Environ. Pollut. **219**, 337–352 (2016b). https://doi.org/10.1016/j.envpol.2016.10.059

Dotse, S.Q., Petra, M.I., Dagar, L., De Silva, L.C.: Application of computational intelligence techniques to forecast daily PM_{10} exceedances in Brunei Darussalam. Atmos. Pollut. Res. **9**(2), 358–368 (2018). https://doi.org/10.1016/j.apr.2017.11.004

Edwards, P.M., Evans, M.J., Furneaux, K.L., Hopkins, J., Ingham, T., Jones, C., Lee, J.D., Lewis, A.C., Moller, S.J., Stone, D., Whalley, L.K., Heard, D.E.: OH reactivity in a South East Asian tropical rainforest during the Oxidant and Particle Photochemical Processes (OP3) project. Atmos. Chem. Phys. **13**, 9497–9514 (2013). https://doi.org/10.5194/acp-13-9497-2013

Fandi, N.F.M., Jalaludin, J., Latif, M.T., Hamid, H.H.A., Awang, M.F.: BTEX exposure assessment and inhalation health risks to traffic policemen in the Klang Valley region, Malaysia. Aerosol Air Qual. Res. **20**, 1922–1937 (2020). https://doi.org/10.4209/aaqr.2019.11.0574

Fujii, Y., Iriana, W., Oda, M., Puriwigati, A., Tohno, S., Lestari, P., Mizohata, A., Huboyo, H.S.: Characteristics of carbonaceous aerosols emitted from peatland fire in Riau, Sumatra, Indonesia. Atmos. Environ. **87**, 164–169 (2014). https://doi.org/10.1016/j.atmosenv.2014.01.037

Fujii, Y., Tohno, S., Ikeda, K., Mahmud, M., Takenaka, N.: A preliminary study on humic-like substances in particulate matter in Malaysia influenced by Indonesian peatland fires. Sci. Total Environ. **753**, 142009 (2021). https://doi.org/10.1016/j.scitotenv.2020.142009

George, S., Chua, M.L., ZheWei, D.Z., Das, R., Bijin, V.A., Connolly, J.E., Lee, K.P., Yung, C.F., Teoh, O.H., Thomas, B.: Personal level exposure and hazard potential of particulate matter during haze and non-haze periods in Singapore. Chemosphere **243**, 125401 (2020). https://doi.org/10.1016/j.chemosphere.2019.125401

Guo, Y., Li, S., Tawatsupa, B., Punnasiri, K., Jaakkola, J.J., Williams, G.: The association between air pollution and mortality in Thailand. Sci. Rep. **4**(1), 5509 (2014). https://doi.org/10.1038/srep05509

Halim, N.D.A., Latif, M.T., Mohamed, A.F., Maulud, K.N.A., Idrus, S., Azhari, A., Othman, M., Sofwan, N.M.: Spatial assessment of land use impact on air quality in mega urban regions, Malaysia. Sustain. Cities Soc. **63**, 102436 (2020). https://doi.org/10.1016/j.scs.2020.102436

Hamid, H.H.A., Latif, M.T., Nadzir, M.S.M., Uning, R., Khan, M.F., Kannan, N.: Ambient BTEX levels over urban, suburban and rural areas in Malaysia. Air Qual. Atmos. Health **12**, 341–351 (2019). https://doi.org/10.1007/s11869-019-00664-1

Harper, K.L., Unger, N.: Global climate forcing driven by altered BVOC fluxes from 1990 to 2010 land cover change in maritime Southeast Asia. Atmos. Chem. Phys. **18**, 16931–16952 (2018). https://doi.org/10.5194/acp-18-16931-2018

Health Effects Institute (HEI).: State of Global Air 2024. Special Report. Health Effects Institute, Boston, MA (2024). https://www.stateofglobalair.org/

Hewitt, C.N., Lee, J.D., MacKenzie, A.R., Barkley, M.P., Carslaw, N., Carver, G.D., Chappell, N.A., Coe, H., Collier, C., Commane, R., Davies, F., Davison, B., DiCarlo, P., Di Marco, C.F., Dorsey, J.R., Edwards, P.M., Evans, M.J., Fowler, D., Furneaux, K.L., Gallagher, M., Guenther, A., Heard, D.E., Helfter, C., Hopkins, J., Ingham, T., Irwin, M., Jones, C., Karunaharan, A., Langford, B., Lewis, A.C., Lim, S.F., MacDonald, S.M., Mahajan, A.S., Malpass, S., McFiggans, G., Mills, G., Misztal, P., Moller, S., Monks, P.S., Nemitz, E., Nicolas-Perea, V., Oetjen, H., Oram, D.E., Palmer, P.I., Phillips, G.J., Pike, R., Plane, J.M.C., Pugh, T., Pyle, J.A., Reeves, C.E., Robinson, N.H., Stewart, D., Stone, D., Whalley, L.K., Yin, X.: Overview: oxidant and particle photochemical processes above a south-east Asian tropical rainforest (the OP3 project): introduction, rationale, location characteristics and tools. Atmos. Chem. Phys. **10**, 169–199 (2010). https://doi.org/10.5194/acp-10-169-2010

Hewitt, C.N., MacKenzie, A.R., Di Carlo, P., Di Marco, C.F., Dorsey, J.R., Evans, M., Fowler, D., Gallagher, M.W., Hopkins, J.R., Jones, C.E., Langford, B., Lee, J.D., Lewis, A.C., Lim,

S.F., McQuaid, J., Misztal, P., Moller, S.J., Monks, P.S., Nemitz, E., Oram, D.E., Owen, S.M., Phillips, G.J., Pugh, T.A.M., Pyle, J.A., Reeves, C.E., Ryder, J., Siong, J., Skiba, U., Stewart, D.J.: Nitrogen management is essential to prevent tropical oil palm plantations from causing ground-level ozone pollution. Proc. Natl. Acad. Sci. **106**, 18447–18451 (2009). https://doi.org/10.1073/pnas.0907541106

Hien, T.T., Huy, D.H., Dominutti, P.A., Chi, N.D.T., Hopkins, J.R., Shaw, M., Forster, G., Mills, G., Le, H.A., Oram, D.: Comprehensive volatile organic compound measurements and their implications for ground-level ozone formation in the two main urban areas of Vietnam. Atmos. Environ. **269**, 118872 (2022). https://doi.org/10.1016/j.atmosenv.2021.118872

Hilario, M.R.A., Bañaga, P.A., Betito, G., Braun, R.A., Cambaliza, M.O., Cruz, M.T., Lorenzo, G.R., MacDonald, A.B., Pabroa, P.C., Simpas, J.B., Stahl, C.: Stubborn aerosol: why particulate mass concentrations do not drop during the wet season in Metro Manila, Philippines. Environ. Sci. Atmos. **2**, 1428–1437 (2022). https://doi.org/10.1039/D2EA00073C

Ho, A.F.W., Ho, J.S., Tan, B.Y.Q., Saffari, S.E.; Yeo, J.W., Sia, C.H., Wang, M., Aik, J., Zheng, H., Morgan, G., San Tam, W.W., Wei, J.S., Ong, M.E.H.: Air quality and the risk of out-of-hospital cardiac arrest in Singapore (PAROS): a time series analysis. Lancet Public Health **7**(11), e932–e941 (2022). https://doi.org/10.1016/S2468-2667(22)00234-1

Ho, A.F.W., Zheng, H., Earnest, A., Cheong, K.H., Pek, P.P., Seok, J.Y., Liu, N., Kwan, Y.H., Tan, J.W.C., Wong, T.H., Hausenloy, D.J.: Time-stratified case crossover study of the association of outdoor ambient air pollution with the risk of acute myocardial infarction in the context of seasonal exposure to the Southeast Asian haze problem. J. Am. Heart Assoc. **8**, e011272 (2019). https://doi.org/10.1161/JAHA.118.011272

Ho, B.Q.: Modeling PM_{10} in Ho Chi Minh City, Vietnam and evaluation of its impact on human health. Sustain. Environ. Res. **27**, 95–102 (2017). https://doi.org/10.1016/j.serj.2017.01.001

Hou, S., Li, W., Yang, L., Chen, G., Zhang, Y., Kuwata, M.: The role of sulfur emission from the petroleum industry on ultrafine particle number concentration in Singapore. Aerosol Air Qual. Res. **23**, 220265 (2023). https://doi.org/10.4209/aaqr.220265

Huy, D.H., Chi, N.D.T., Nam, N.X.T., Hien, T.T.: Commuter exposures to in-transit PM in an urban city dominated by motorcycle: a case study in Vietnam. Atmos. Pollut. Res. **13**(3), 101351 (2022). https://doi.org/10.1016/j.apr.2022.101351

Huy, L.N., Oanh, N.T.K., Htut, T.T., Hlaing, O.M.T.: Emission inventory for on-road traffic fleets in Greater Yangon, Myanmar. Atmos. Pollut. Res. **11**(4), 702–713 (2020). https://doi.org/10.1016/j.apr.2019.12.021

Institute for Health Metric and Evaluation (IHME). Global Burden of Diseases 2019. University of Washington. https://www.healthdata.org/gbd/2019 (2020)

Jayarathne, T., Stockwell, C.E., Gilbert, A.A., Daugherty, K., Cochrane, M.A., Ryan, K.C., Putra, E.I., Saharjo, B.H., Nurhayati, A.D., Albar, I., Yokelson, R.J., Stone, E.'A.: Chemical character-ization of fine particulate matter emitted by peat fires in Central Kalimantan, Indonesia, during the 2015 El Niño. Atmos. Chem. Phys. **18**, 2585–2600 (2018). https://doi.org/10.5194/acp-18-2585-2018

Kasthuriarachchi, N.Y., Rivellini, L.H., Adam, M.G., Lee, A.K.: Light absorbing properties of primary and secondary brown carbon in a tropical urban environment. Environ. Sci. Technol. **54**(17), 10808–10819 (2020). https://doi.org/10.1021/acs.est.0c02414

Kecorius, S., Madueño, L., Vallar, E., Alas, H., Betito, G., Birmili, W., Cambaliza, M.O., Catipay, G., Gonzaga-Cayetano, M., Galvez, M.C., Lorenzo, G., Müller, T., Simpas, J.B., Tamayo, E.G., Wiedensohler, A.: Aerosol particle mixing state, refractory particle number size distributions and emission factors in a polluted urban environment: Case study of Metro Manila, Philippines. Atmos. Environ. **170**, 169–183 (2017). https://doi.org/10.1016/j.atmosenv.2017.09.037

Khan, M.F., Sulong, N.A., Latif, M.T., Nadzir, M.S.M., Amil, N., Hussain, D.F.M., Lee, V., Hosaini, P.N., Shaharom, S., Yusoff, N.A.Y.M., Hoque, H.M.S., Chung, J.X., Sahani, M., Tahir, N.M., Juneng, L., Maulud, K.N.A., Abdullah, S.M.S., Fujii, Y., Toohno, S., Mizohata, A.: Compre-hensive assessment of $PM_{2.5}$ physicochemical properties during the Southeast Asia dry season (southwest monsoon). J. Geophys. Res. Atmos. **121**(24), 14589–14611 (2016). https://doi.org/10.1002/2016JD025894

Kongpran, J., Kliengchuay, W., Niampradit, S., Sahanavin, N., Siriratruengsuk, W., Tantrakarnapa, K.: The health risks of airborne polycyclic aromatic hydrocarbons (PAHs): upper North Thailand. Geohealth **5**(4), e2020GH000352 (2021). https://doi.org/10.1029/2020GH000352

Kusumaningtyas, S.D.A., Tonokura, K., Aldrian, E., Giles, D.M., Holben, B.N., Gunawan, D., Lestari, P., Iriana, W.: Aerosols optical and radiative properties in Indonesia based on AERONET version 3. Atmos. Environ. **282**, 119174 (2022). https://doi.org/10.1016/j.atmosenv.2022.119174

Langford, B., Misztal, P.K., Nemitz, E., Davison, B., Helfter, C., Pugh, T.A.M., MacKenzie, A.R., Lim, S.F., Hewitt, C.N.: Fluxes and concentrations of volatile organic compounds from a South-East Asian tropical rainforest. Atmos. Chem. Phys. **10**, 8391–8412 (2010). https://doi.org/10.5194/acp-10-8391-2010

Latif, M.T., Dominick, D., Ahamad, F., Khan, M.F., Juneng, L., Hamzah, F.M., Nadzir, M.S.M.: Long term assessment of air quality from a background station on the Malaysian Peninsula. Sci. Total Environ. **482–483**, 336–348 (2014). https://doi.org/10.1016/j.scitotenv.2014.02.132

Lestari, P., Arrohman, M.K., Damayanti, S., Klimont, Z.: Emissions and spatial distribution of air pollutants from anthropogenic sources in Jakarta. Atmos. Pollut. Res. **13**(9), 101521 (2022). https://doi.org/10.1016/j.apr.2022.101521

Lorenzo, G.R., Arellano, A.F., Cambaliza, M.O., Castro, C., Cruz, M.T., Di Girolamo, L., Gacal, G.F., Hilario, M.R.A., Lagrosas, N., Ong, H.J., Simpas, J.B., Uy, S.N., Sorooshian, A.: An emerging aerosol climatology via remote sensing over Metro Manila, the Philippines. Atmos. Chem. Phys. **23**, 10579–10608 (2023). https://doi.org/10.5194/acp-23-10579-2023

Lorenzo, G.R., Bañaga, P.A., Cambaliza, M.O., Cruz, M.T., AzadiAghdam, M., Arellano, A., Betito, G., Braun, R., Corral, A.F., Dadashazar, H., Edwards, E.-L., Eloranta, E., Holz, R., Leung, G., Ma, L., MacDonald, A.B., Reid, J.S., Simpas, J.B., Stahl, C., Visaga, S.M., Sorooshian, A.: Measurement report: Firework impacts on air quality in Metro Manila, Philippines, during the 2019 New Year revelry. Atmos. Chem. Phys. **21**, 6155–6173 (2021). https://doi.org/10.5194/acp-21-6155-2021

Lu, L., Ng, V.Y.Z., Tan, M.Z.H., Kasthuriarachchi, N.Y., Rivellini, L.H., Yue, Q.T., Ang, L., Viera, M., Bay, B.H., Seow, W.J., Lee, A.K.Y.: Particle-bound reactive oxygen species in cooking emissions: aging effects and cytotoxicity. Atmos. Environ. **319**, 120309 (2024). https://doi.org/10.1016/j.atmosenv.2023.120309

Luong, L.M., Phung, D., Sly, P.D., Morawska, L., Thai, P.K.: The association between particulate air pollution and respiratory admissions among young children in Hanoi, Vietnam. Sci. Total Environ. **578**, 249–255 (2017). https://doi.org/10.1016/j.scitotenv.2016.08.012

Luong, L.T.M., Dang, T.N., Huong, N.T.T., Phung, D., Tran, L.K., Dung, D.V., Thai, P.K.: Particulate air pollution in Ho Chi Minh city and risk of hospital admission for acute lower respiratory infection (ALRI) among young children. Environ. Pollut. **25**, 113424 (2020). https://doi.org/10.1016/j.envpol.2019.113424

Ly, B.T., Kajii, Y., Shoji, K., Van, D.A., Nghiem, T.D., Sakamoto, Y.: Characteristics of roadside volatile organic compounds in an urban area dominated by gasoline vehicles, a case study in Hanoi. Chemosphere **254**, 126749 (2020). https://doi.org/10.1016/j.chemosphere.2020.126749

Maciaszek, K., Gillies, S., Kawichai, S., Prapamontol, T., Santijitpakdee, T., Kliengchuay, W., Sahanavin, N., Mueller, W., Vardoulakis, S., Samutrtai, P., Cherrie, J.W.: In vitro assessment of the pulmonary toxicity of particulate matter emitted during haze events in Chiang Mai, Thailand via investigation of macrophage responses. Environ. Res. Health **1**(2), 025002 (2023). https://doi.org/10.1088/2752-5309/ac9748

Madueño, L., Kecorius, S., Birmili, W., Müller, T., Simpas, J., Vallar, E., Galvez, M.C., Cayetano, M., Wiedensohler, A.: Aerosol particle and black carbon emission factors of vehicular fleet in Manila, Philippines. Atmosphere **10**(10), 603 (2019). https://doi.org/10.3390/atmos10100603

Madueño, L., Kecorius, S., Löndahl, J., Schnelle-Kreis, J., Wiedensohler, A., Pöhlker, M.: A novel in-situ method to determine the respiratory tract deposition of carbonaceous particles reveals dangers of public commuting in highly polluted megacity. Part. Fibre Toxicol. **19**(1), 61 (2022). https://doi.org/10.1186/s12989-022-00501-x

Mahiyuddin, W.R.W., Sahani, M., Aripin, R., Latif, M.T., Thach, T.Q., Wong, C.M.: Short-term effects of daily air pollution on mortality. Atmos. Environ. **65**, 69–79 (2013). https://doi.org/10. 1016/j.atmosenv.2012.10.019

Makkonen, U., Vestenius, M., Huy, L.N., Anh, N.T.N., Linh, V., Pham, T., Ha, M.P., Nguyen, H., Aurela, M., Hellén, H., Loven, K., Koznetsov, R., Kyllönen, K., Teinilä, K., Oanh, N.T.K.: Chemical composition and potential sources of PM$_{2.5}$ in Hanoi. Atmos. Environ. **299**, 119650 (2023). https://doi.org/10.1016/j.atmosenv.2023.119650

Misztal, P.K., Nemitz, E., Langford, B., Di Marco, C.F., Phillips, G.J., Hewitt, C.N., MacKenzie, A.R., Owen, S.M., Fowler, D., Heal, M.R., Cape, J.N.: Direct ecosystem fluxes of volatile organic compounds from oil palms in South-East Asia. Atmos. Chem. Phys. **11**, 8995–9017 (2011). https://doi.org/10.5194/acp-11-8995-201

Nghiem, T.D., Mac, D.H., Nguyen, A.D., Lê, N.C.: An integrated approach for analyzing air quality monitoring data: a case study in Hanoi, Vietnam. Air Qual. Atmos. Health **14**, 7–18 (2021). https://doi.org/10.1007/s11869-020-00907-6

Nghiem, T.D., Nguyen, Y.L.T., Le, A.T., Bui, N.D., Pham, H.T.: Development of the specific emission factors for buses in Hanoi, Vietnam. Environ. Sci. Pollut. Res. **26**, 24176–24189 (2019). https://doi.org/10.1007/s11356-019-05634-9

Nhung, N.T.T., Schindler, C., Dien, T.M., Probst-Hensch, N., Künzli, N.: Association of ambient air pollution with lengths of hospital stay for Hanoi children with acute lower-respiratory infection, 2007–2016. Environ. Pollut. **247**, 752–762 (2019). https://doi.org/10.1016/j.envpol.2019.01.115

Oanh, N.T.K., Tipayarom, A., Bich, T.L., Tipayarom, D., Simpson, C.D., Hardie, D., Liu, L.J.S.: Characterization of gaseous and semi-volatile organic compounds emitted from field burning of rice straw. Atmos. Environ. **119**, 182–191 (2015). https://doi.org/10.1016/j.atmosenv.2015. 08.005

Othman, J., Sahani, M., Mahmud, M., Ahmad, M.K.S.: Transboundary smoke haze pollution in Malaysia: inpatient health impacts and economic valuation. Environ. Pollut. **189**, 194–201 (2014). https://doi.org/10.1016/j.envpol.2014.03.010

Othman, M., Latif, M.T., Hamid, H.H.A., Uning, R., Khumsaeng, T., Phairuang, W., Daud, Z., Idris, J., Sofwan, N.M., Lung, S.C.C.: Spatial–temporal variability and health impact of particulate matter during a 2019–2020 biomass burning event in Southeast Asia. Sci. Rep. **12**(1), 7630 (2022). https://doi.org/10.1038/s41598-022-11409-z

Pani, S.K., Lin, N.H., Chantara, S., Wang, S.H., Khamkaew, C., Prapamontol, T., Janjai, S.: Radiative response of biomass-burning aerosols over an urban atmosphere in northern peninsular Southeast Asia. Sci. Total Environ. **633**, 892–911 (2018). https://doi.org/10.1016/j.scitotenv.2018.03.204

Pavagadhi, S., Betha, R., Venkatesan, S., Balasubramanian, R., Hande, M.P.: Physicochemical and toxicological characteristics of urban aerosols during a recent Indonesian biomass burning episode. Environ. Sci. Pollut. Res. **20**, 2569–2578 (2013). https://doi.org/10.1007/s11356-012- 1157-9

Phairuang, W., Inerb, M., Furuuchi, M., Hata, M., Tekasakul, S., Tekasakul, P.: Size-fractionated carbonaceous aerosols down to PM$_{0.1}$ in southern Thailand: local and long-range transport effects. Environ. Pollut. **260**, 114031 (2020). https://doi.org/10.1016/j.envpol.2020.114031

Pham, C.T., Ly, B.T., Nghiem, T.D., Pham, T.H.P., Minh, N.T., Tang, N., Hayakawa, K., Toriba, A.: Emission factors of selected air pollutants from rice straw burning in Hanoi, Vietnam. Air Qual. Atmos. Health **14**, 1757–1771 (2021). https://doi.org/10.1007/s11869-021-01050-6

Phosri, A., Ueda, K., Phung, V.L.H., Tawatsupa, B., Honda, A., Takano, H.: Effects of ambient air pollution on daily hospital admissions for respiratory and cardiovascular diseases in Bangkok, Thailand. Sci. Total Environ. **651**, 1144–1153 (2019). https://doi.org/10.1016/j.scitotenv.2018. 09.183

Phuc, N.H., Oanh, N.T.K.: Determining factors for levels of volatile organic compounds measured in different microenvironments of a heavy traffic urban area. Sci. Total Environ. **627**, 290–303 (2018). https://doi.org/10.1016/j.scitotenv.2018.01.216

Phung, D., Hien, T.T., Linh, H.N., Luong, L.M.T., Morawska, L., Chu C., Bindh, N.D., Thai. P.K.: Air pollution and risk of respiratory and cardiovascular hospitalizations in the most populous city in Vietnam. Sci. Total Environ. **557–558**, 322–330 (2016). https://doi.org/10.1016/j.scitot env.2016.03.070

Phuong, P.T.H., Nghiem, T.D., Thao, P.T.M., Nguyen, T.D.: Emission factors of selected air pollutants from rice straw open burning in the Mekong Delta of Vietnam. Atmos. Pollut. Res. **13**(3), 101353 (2022). https://doi.org/10.1016/j.apr.2022.101353

Pinichka, C., Makka, N., Sukkumnoed, D., Chariyalertsak, S., Inchai, P., Bundhamcharoen, K.: Burden of disease attributed to ambient air pollution in Thailand: a GIS-based approach. PLoS ONE **12**(12), e0189909 (2017). https://doi.org/10.1371/journal.pone.0189909

Pongpiachan, S., Tipmanee, D., Khumsup, C., Kittikoon, I., Hirunyatrakul, P.: Assessing risks to adults and preschool children posed by $PM_{2.5}$-bound polycyclic aromatic hydrocarbons (PAHs) during a biomass burning episode in Northern Thailand. Sci. Total Environ. **508**, 435–444 (2015). https://doi.org/10.1016/j.scitotenv.2014.12.019

Promsiri, P., Tekasakul, S., Thongyen, T., Suwattiga, P., Morris, J., Latif, M.T., Tekasakul, P., Dejchanchaiwong, R.: Transboundary haze from peatland fires and local source-derived $PM_{2.5}$ in Southern Thailand. Atmos. Environ. **294**, 119512 (2023). https://doi.org/10.1016/j.atmosenv. 2022.119512

Quang, T.N., Hue, N.T., Tran, L.K., Phi, T.H., Morawska, L., Thai, P.K.: Motorcyclists have much higher exposure to black carbon compared to other commuters in traffic of Hanoi, Vietnam. Atmos. Environ. **245**, 118029 (2021). https://doi.org/10.1016/j.atmosenv.2020.118029

Rivellini, L.-H., Adam, M.G., Kasthuriarachchi, N., Lee, A.K.Y.: Characterization of carbonaceous aerosols in Singapore: insight from black carbon fragments and trace metal ions detected by a soot particle aerosol mass spectrometer. Atmos. Chem. Phys. **20**(10), 5977–5993 (2020). https:// doi.org/10.5194/acp-20-5977-2020

Samsuddin, N.A.C., Khan, M.F., Maulud, K.N.A., Hamid, A.H., Munna, F.T., Ab Rahim, M.A., Latif, M.T., Akhtaruzzaman, M.: Local and transboundary factors' impacts on trace gases and aerosol during haze episode in 2015 El Niño in Malaysia. Sci. Total Environ. **630**, 1502–1514 (2018). https://doi.org/10.1016/j.scitotenv.2018.02.289

Santoso, M., Lestiani, D.D., Kurniawati, S., Damastuti, E., Kusmartini, I., Atmodjo, D.P.D., Sari, D.K., Hopke, P.K., Mukhtar, R., Muhtarom, T., Tjahyadi, A., Parian, S., Kholik, N., Sutrisno, D.A., Wahyudi, D., Sitorus, T.D., Djamilus, J., Riadi, A., Supriyanto, J., Dahyar, N., Sondakh, S., Hogendorp, K., Wahyuni, N., Bejawan, I.G., Suprayadi, L.S.: Assessment of urban air quality in Indonesia. Aerosol Air Qual. Res. **20**, 2142–2158 (2020). https://doi.org/10.4209/aaqr.2019. 09.0451

Seinfeld, J.H., Pandis, S.N.: Atmospheric Chemistry and Physics: from Air Pollution to Climate Change, 3rd edn. Wiley (2016)

Shi, Y., Matsunaga, T., Yamaguchi, Y., Li, Z., Gu, X., Chen, X.: Long-term trends and spatial patterns of satellite-retrieved $PM_{2.5}$ concentrations in South and Southeast Asia from 1999 to 2014. Sci. Total Environ. **615**, 177–186 (2018). https://doi.org/10.1016/j.scitotenv.2017.09.241

Silva, S.J., Heald, C.L., Geddes, J.A., Austin, K.G., Kasibhatla, P.S., Marlier, M.E.: Impacts of current and projected oil palm plantation expansion on air quality over Southeast Asia. Atmos. Chem. Phys. **16**, 10621–10635 (2016). https://doi.org/10.5194/acp-16-10621-2016

Siregar, S., Idiawati, N., Lestari, P., Berekute, A.K., Pan, W.C., Yu, K.P.: Chemical composition, source appointment and health risk of $PM_{2.5}$ and $PM_{2.5-10}$ during forest and peatland fires in Riau, Indonesia. Aerosol Air Qual. Res. **22**(9), 220015 (2022). https://doi.org/10.4209/aaqr. 220015

Sirithian, D., Thepanondh, S., Sattler, M.L., Laowagul, W.: Emissions of volatile organic compounds from maize residue open burning in the northern region of Thailand. Atmos. Environ. **176**, 179–187 (2018). https://doi.org/10.1016/j.atmosenv.2017.12.032

Smith, T.E.L., Evers, S., Yule, C.M., Gan, J.Y.: In situ tropical peatland fire emission factors and their variability, as determined by field measurements in peninsula Malaysia. Global Biogeochem. Cycles **32**, 18–31 (2018). https://doi.org/10.1002/2017gb005709

Sofwan, N.M., Latif, M.T.: Characteristics of the real-driving emissions from gasoline passenger vehicles in the Kuala Lumpur urban environment. Atmos. Pollut. Res. **12**(1), 306–315 (2021). https://doi.org/10.1016/j.apr.2020.09.004

Sothea, K., Oanh, N.T.K.: Characterization of emissions from diesel backup generators in Cambodia. Atmos. Pollut. Res. **10**(2), 345–354 (2019). https://doi.org/10.1016/j.apr.2018.09.001

Stahl, C., Cruz, M.T., Bañaga, P.A., Betito, G., Braun, R.A., Aghdam, M.A., Cambaliza, M.O., Lorenzo, G.R., MacDonald, A.B., Hilario, M.R.A., Pabroa, P.C.: Sources and characteristics of size-resolved particulate organic acids and methanesulfonate in a coastal megacity: Manila, Philippines. Atmos. Chem. Phys. **20**(24), 15907–15935 (2020). https://doi.org/10.5194/acp-20-15907-2020

Stockwell, C.E., Jayarathne, T., Cochrane, M.A., Ryan, K.C., Putra, E.I., Saharjo, B.H., Nurhayati, A.D., Albar, I., Blake, D.R., Simpson, I.J., Stone, E.A., Yokelson, R.J.: Field measurements of trace gases and aerosols emitted by peat fires in Central Kalimantan, Indonesia, during the 2015 El Niño. Atmos. Chem. Phys. **16**, 11711–11732 (2016). https://doi.org/10.5194/acp-16-11711-2016

Stone, D., Whalley, L.K., Heard, D.E.: Tropospheric OH and HO_2 radicals: field measurements and model comparisons. Chem. Soc. Rev. **41**, 6348–6404 (2012). https://doi.org/10.1039/C2CS35 140D

Sukkhum, S., Lim, A., Ingviya, T., Saelim, R.: Seasonal patterns and trends of air pollution in the upper northern Thailand from 2004 to 2018. Aerosol Air Qual. Res. **22**(5), 210318 (2022). https://doi.org/10.4209/aaqr.210318

Sulong, N.A., Latif, M.T., Khan, M.F., Amil, N., Ashfold, M.J., Wahab, M.I.A., Chan, K.M., Sahani, M.: Source apportionment and health risk assessment among specific age groups during haze and non-haze episodes in Kuala Lumpur, Malaysia. Sci. Total Environ. **601**, 556–570 (2017). https://doi.org/10.1016/j.scitotenv.2017.05.153

Tan, B.Y.Q., Ho, J.S.Y., Ho, A.F.W., Pek, P.P., Leow, A.S.T., Raju, Y., Sia, C.H., Yeo, L.L.L., Sharma, V.K., Ong, M.E.H., Aik, J., Zheng, H.: Ambient air pollution and acute ischemic stroke-effect modification by atrial fibrillation. J. Clin. Med. **11**(18), 5429 (2022). https://doi.org/10.3390/jcm11185429

Tan, S.H., Roth, M., Velasco, E.: Particle exposure and inhaled dose during commuting in Singapore. Atmos. Environ. **170**, 245–258 (2017). https://doi.org/10.1016/j.atmosenv.2017.09.056

Tham, J., Sarkar, S., Jia, S., Reid, J.S., Mishra, S., Sudiana, I.M., Swarup, S., Ong, C.N., Liya, E.Y.: Impacts of peat-forest smoke on urban $PM_{2.5}$ in the Maritime Continent during 2012–2015: carbonaceous profiles and indicators. Environ. Pollut. **248**, 496–505 (2019). https://doi.org/10.1016/j.envpol.2019.02.049

Thepnuan, D., Chantara, S., Lee, C.T., Lin, N.H., Tsai, Y.I.: Molecular markers for biomass burning associated with the characterization of $PM_{2.5}$ and component sources during dry season haze episodes in Upper South East Asia. Sci. Total Environ. **658**, 708–722 (2019). https://doi.org/10.1016/j.scitotenv.2018.12.201

Tunsaringkarn, T., Prueksasit, T., Morknoy, D., Semathong, S., Rungsiyothin, A., Zapaung, K.: Ambient air's volatile organic compounds and potential ozone formation in the urban area, Bangkok, Thailand. J. Environ. Occup. Health **3**(3), 130–135 (2014). https://doi.org/10.5455/jeos.20140903015449

Uttamang, P., Aneja, V.P., Hanna, A.F.: Assessment of gaseous criteria pollutants in the Bangkok Metropolitan Region, Thailand. Atmos. Chem. Phys. **18**, 12581–12593 (2018). https://doi.org/10.5194/acp-18-12581-2018

Velasco, E., Rastan, S.: Air quality in Singapore during the 2013 smoke-haze episode over the Strait of Malacca: lessons learned. Sustain. Cities Soc. **17**, 122–131 (2015). https://doi.org/10.1016/j.scs.2015.04.006

Velasco, E., Segovia E.: Effectiveness of equipping bus stop shelters with cooling and filtering systems in a city with tropical climate. Smart Sustain. Built Environ **13**(5), 1330-1345 (2024). https://doi.org/10.1108/SASBE-03-2023-0063

Velasco, E., Tan, S.H.: Exposure while sitting at bus stops of hot and humid Singapore. Atmos. Environ. **142**, 251–263 (2016). https://doi.org/10.1016/j.atmosenv.2016.07.054

Velasco, E., Ha, H.H., Pham, A.D., Rastan S.: Effectiveness of wearing face masks against traffic particles on the streets of Ho Chi Minh City, Vietnam. Environ. Sci. Atmos. **2**, 1450-1468 (2022). https://doi.org/10.1039/D2EA00071G

Velasco, E., Ho, K.J.J., Ziegler, A.: Commuter exposure to black carbon, carbon monoxide, and noise in the mass transport *khlong* boats of Bangkok, Thailand. Transport Res. Part D Transport Environ. **21**, 62–65 (2013). https://doi.org/10.1016/j.trd.2013.02.010

Velasco, E., Segovia, E.: Determining a commuters' exposure to particle and noise pollution on double-decker buses. Aerosol Air Qual. Res. **21**(12), 210165 (2021). https://doi.org/10.4209/aaqr.210165

Wang, S.H., Welton, E.J., Holben, B.N., Tsay, S.C., Lin, N.H., Giles, D., Stewart, S.A., Janjai, S., Nguyen, X.A., Hsiao, T.C., Chen, W.N., Lin, T.H., Buntoung, S., Chantara, S., Wiriya, W.: Vertical distribution and columnar optical properties of springtime biomass-burning aerosols over Northern Indochina during 2014 7-SEAS campaign. Aerosol Air Qual. Res. **15**, 2037–2050 (2015). https://doi.org/10.4209/aaqr.2015.05.0310

Warwick, N.J., Archibald, A.T., Ashworth, K., Dorsey, J., Edwards, P.M., Heard, D.E., Langford, B., Lee, J., Misztal, P.K., Whalley, L.K., Pyle, J.A.: A global model study of the impact of land-use change in Borneo on atmospheric composition. Atmos. Chem. Phys. **13**, 9183–9194 (2013). https://doi.org/10.5194/acp-13-9183-2013

Wiggins, E.B., Czimczik, C.I., Santos, G.M., Chen, Y., Xu, X., Holden, S.R., Randerson, J.T., Harvey, C.F., Kai, F.M., Yu, L.E.: Smoke radiocarbon measurements from Indonesian fires provide evidence for burning of millennia-aged peat. Proc. Natl. Acad. Sci. **115**, 12419–12424 (2018). https://doi.org/10.1073/pnas.1806003115

World Bank: The global health cost of $PM_{2.5}$ air pollution: a case for action beyond 2021. In: International Development in Focus. World Bank Washington, DC. https://openknowledge.worldbank.org/handle/10986/36501 (2022)

Yang, Y., Shao, M., Wang, X., Nölscher, A.C., Kessel, S., Guenther, A., Williams, J.: Towards a quantitative understanding of total OH reactivity: a review. Atmos. Environ. **134**, 147–161 (2016). https://doi.org/10.1016/j.atmosenv.2016.03.010

Yokelson, R.J., Saharjo, B.H., Stockwell, C.E., Putra, E.I., Jayarathne, T., Akbar, A., Albar, I., Blake, D.R., Graham, L.L.B., Kurniawan, A., Meinardi, S., Ningrum, D., Nurhayati, A.D., Saad, A., Sakuntaladewi, N., Setianto, E., Simpson, I.J., Stone, E.A., Sutikno, S., Thomas, A., Ryan, K.C., Cochrane, M.A.: Tropical peat fire emissions: 2019 field measurements in Sumatra and Borneo and synthesis with previous studies. Atmos. Chem. Phys. **22**, 10173–10194 (2022). https://doi.org/10.5194/acp-22-10173-2022

Yorifuji, T., Bae, S., Kashima, S., Tsuda, T., Doi, H., Honda, Y., Kim, H., Hong, Y.C.: Health impact assessment of PM_{10} and $PM_{2.5}$ in 27 Southeast and East Asian cities. J. Occup. Environ. Med. **57**(7), 751–756 (2015). https://doi.org/10.1097/JOM.0000000000000485

Yu, W., Sulistyoningrum, D.C., Gasevic, D., Xu, R., Julia, M., Murni, I.K., Chen, Z., Lu, P., Guo, Y., Li, S.: Long-term exposure to $PM_{2.5}$ and fasting plasma glucose in non-diabetic adolescents in Yogyakarta, Indonesia. Environ. Pollut. **257**, 113423 (2020). https://doi.org/10.1016/j.envpol.2019.113423

Zhang, N., Maung, M.W., Win, M.S., Feng, J., Yao, X.: Carbonaceous aerosol and inorganic ions of $PM_{2.5}$ in Yangon and Mandalay of Myanmar: seasonal and spatial variations in composition and sources. Atmos. Pollut. Res. **13**(6), 101444 (2022). https://doi.org/10.1016/j.apr.2022.101444

Zong, Y., Botero, M.L., Liya, E.Y., Kraft, M.: Size spectra and source apportionment of fine particulates in tropical urban environment during southwest monsoon season. Environ. Pollut. **244**, 477–485 (2019). https://doi.org/10.1016/j.envpol.2018.09.124

Zulkepli, N.F.S., Noorani, M.S.M., Razak, F.A., Ismail, M., Alias, M.A.: Topological characterization of haze episodes using persistent homology. Aerosol Air Qual. Res. **19**, 1614–1624 (2019). https://doi.org/10.4209/aaqr.2018.08.0315

Zulkifli, M.F.H., Hawari, N.S.S.L., Latif, M.T., Abd Hamid, H.H., Mohtar, A.A.A., Idris, W.M.R.W., Mustaffa, N.I.H., Juneng, L.: Volatile organic compounds and their contribution to ground-level ozone formation in a tropical urban environment. Chemosphere **302**, 134852 (2022). https://doi.org/10.1016/j.chemosphere.2022.134852

Chapter 10
Challenges and Recommendations

Abstract This chapter summarizes the challenges that Southeast Asia's air quality management faces nowadays. It discusses how the region's eleven nations could work together to achieve clean air by improving air quality monitoring, building emission inventories to run air quality models, developing air quality models for both research and regulatory purposes, using available global air quality models to develop forecast modeling systems tailored to local conditions at operational scales for precautionary purposes, taking advantage of advances in satellite remote sensing to fill data gaps in regions not yet monitored at ground level, better understanding the connections between air quality and climate change, and promoting research and collaborative intensive field campaigns as a means of improving scientific knowledge. The authors do not intend to propose a roadmap to improve air quality management. Such a task corresponds to local decision-makers, air quality managers, practitioners, and citizens. The recommendations presented here are based on the findings that emerged from the literature review conducted for this monograph, lessons learned in other parts of the world, proposals developed to improve air quality management in low- and middle-income countries, and the authors' own experience.

Keywords Air quality monitoring · Emission inventories · Air quality modeling · Satellite observations · Intensive field campaigns · Climate change

The lack of air quality data in Southeast Asia is critical. The number of regulatory-grade air quality monitoring stations is insufficient to inform about the air pollution to which the more than 650-million residents of this part of the world are exposed. Most of the stations are located in large urban centers, leaving most of the region, including entire countries, without air quality information. There is little data on air pollutants emission. The emission datasets that are available are the result of international initiatives aimed at compiling emission inventories for global and regional air quality assessments. Similarly, the application of air quality models and the use of satellite resources in support of air quality management remains limited. Only a few groups of researchers have used air quality models to answer specific questions on the impact caused by particular emission sources, such as agricultural waste burning, on ambient air pollution. There is insufficient modeling capacity for air quality forecasting in place, despite the fact that every time the region's sky is blanketed for days as a result of massive wildfires caused by aggressive deforestation and agricultural practices, representatives from all countries agree on the need for such a forecast modeling system being implemented.

In terms of scientific research, valuable efforts have been made to improve local knowledge on some issues related to air quality in the region. They are, however, insufficient. Much of the research has focused on particle pollution caused by biomass burning, which is often the source of the most severe episodes of poor air quality. Most of the research has concentrated on seven cities in five countries (Kuala Lumpur in Malaysia, Manila in the Philippines, Bangkok and Chiang Mai in Thailand, Hanoi and Ho Chi Minh City in Vietnam, and the city-sate of Singapore), with little or no research done in the other countries. And, with the exception of Thailand, very little research has been conducted outside of urban areas in the former countries.

Because of the precarious state of air quality management in the region, it is impossible to know the true extent of the threat that is posed by air pollution, making it impossible to take corrective actions. The growing trend of air pollution in the region and corresponding adverse effects on public health and associated economic blow reported by international assessments such as the State of Global Air 2024 (HEI 2024), the Global Burden of Diseases 2019 (IHME 2020), and the Global Health Cost of $PM_{2.5}$ Air Pollution (World Bank 2022) are useful indicators of the magnitude of the problem, they can even be seen as the most up-to-date and complete source of information, but they are still mere indicators. Their accuracy is jeopardized by a lack of information on the ground (i.e., air quality data, emission inventories, and demographic and economic statistics), as well as the representativeness of the models and methodologies used for Southeast Asia conditions. In regions with limited data, the results of these assessments are often conservative to avoid raising unfounded concern. As a result, it should not be surprising if the size of the problem is even bigger.

This section summarizes the challenges that Southeast Asia's air quality management faces due to a lack of data on the matter. It discusses how the region's eleven nations could work together to achieve clean air by improving air quality monitoring and data accessibility, building emission inventories that can be used to run air quality models, developing air quality models for both research and regulatory purposes, using already available global and regional air quality models to develop air quality forecast modeling systems tailored to local conditions at operational scales for precautionary purposes, taking advantage of recent advances in satellite remote sensing to fill air quality data gaps in regions not yet monitored at ground level and to identify plumes from major emission sources, and promoting research and collaborative intensive field campaigns as a means of improving scientific knowledge to address air pollution.

The recommendations presented here are based on the findings that emerged from the literature review conducted for this work, lessons learned in other parts of the world (e.g., Molina et al. 2019; Hidy et al. 2011), proposals developed to improve air quality management in low- and middle-income countries (e.g., Gani et al. 2022; Pinder et al. 2019), and the authors' experience on the subject.

The authors do not intend to propose a roadmap to improve air quality management. Such a task corresponds to local decision-makers, air quality managers, and citizens, who should work together to find collective solutions at the local and regional scales. The World Health Organization makes available a set of tools for assessing and addressing air pollution and health that could be used as a starting point for such an endeavor (WHO 2023). These tools, which have been compiled in a common online repository, aim to improve countries' ability to report on air quality, recognizing it as a key component for the development of a clean air agenda.

The Opportunity Score proposed by the Energy Policy Institute at the University of Chicago (EPCI) is a tool that countries and provinces with limited or no air quality monitoring infrastructure could use to assess the impact of improving air quality monitoring on clean air policy (Hasenkopf et al. 2023). It focuses on $PM_{2.5}$ and provides a metric for determining where new monitoring stations would be most useful in filling data gaps to address air pollution.

10.1 Air Quality Monitoring

Reliable data on key pollutant concentrations at the ambient level is critical for understanding and taking corrective action to improve air quality. Air quality monitoring allows to build air pollution trends and objectively assess the impact of economic growth and urbanization, as well as the outcomes of environmental policies and control measures. Despite the importance of measuring the abundance of pollutants in the atmospheric environment to protect public health, Lao PDR, Myanmar, and Timor-Leste do not have any air quality monitoring system in place; the other eight countries of Southeast Asia monitor air quality to some extent, primarily in capital cities and major urban areas (see Table 5.2). There is little air quality data in areas of ecological value, agricultural extensions (important for food security), and small towns. With the exception of Brunei Darussalam and Singapore, the coverage of monitoring stations is insufficient to address air pollution throughout the region (Fig. 5.2). Countries with monitoring stations use pollutants readings to calculate air quality indices, which inform citizens about the air pollution level to which they are exposed. According to local legislation, some countries also publish rolling average concentrations every hour, but not time-resolved hourly concentration values, with the exception of $PM_{2.5}$ and NO_2 in some cases. In general, public access to archived and historic records is limited, and databases are inaccessible. Concerns about data misuse or misinterpretation prevent local governments from openly sharing information; additionally, it is not an official mandate for environmental agencies.

All Southeast Asian nations must work together to address the lack of air quality monitoring across the region. Inequality in access to air quality data is evident. This will require coordination among countries, including technical and economic participation. Most advanced nations on the matter should provide technical assistance, while those with strong economies should provide financial aid. Efforts should first focus on providing monitoring stations equipped with regulatory-grade monitors in major urban centers, followed by smaller cities and natural or rural areas. The location of monitoring stations and the pollutant species to be measured (it may not be necessary to measure all criteria pollutants in each station) should allow authorities to meet the following objectives:

- Assess air pollution levels in relation to regulatory requirements and health-based targets.
- Have a warning system in place to avoid/reduce exposure to high levels of air pollution.
- Track progress and compliance with air quality regulations (i.e., air quality standards).
- Enforce control measure policies and evaluate their effectiveness.
- Identify pollutant sources and local hot-spots of air pollution.
- Evaluate the impact on public health.
- Raise public awareness.

It would be useful to have protocols for the design and operation of air quality monitoring networks based on air quality management requirements. These protocols would act as guidelines for both new and existing networks to collect spatially and temporally representative data that can be used for exposure assessment, air quality management, and regulatory compliance (i.e., air quality standards). These protocols would also help in the configuration of networks that take into account aspects of sustainability and environmental justice. Accountability is another aspect to consider, the protocols must lay the foundation for this in order to ensure that the measures to improve air quality are actually being implemented.

Coordination will also be required to harmonize operation, maintenance, and quality assurance and quality control (QA/QC) in accordance with established protocols. This will avoid, or at least reduce, disparities in data quality. Periodic technical audits and performance evaluations should be conducted by specialized personnel or a third party not involved in local monitoring programs (e.g., a team of air quality monitoring experts from another region of the world) recognized by all Southeast Asian nations for such an endeavor. These audits would allow for the identification of capacities and deficiencies at each stage of the monitoring process, as well as the development of objective and realistic work plans to ensure adequate operation in the short term.

Similarly, the entire region would benefit if data from all monitoring stations, in addition to local repositories, were stored and harmonized in a common repository. To avoid poor quality data being included, a validation system and a set of QA/QC procedures compatible with those established in each country would be required.

A regional coordination and operation center would assist in all air quality monitoring activities. This center would work closely with local agencies to help them improve their operational performance. It would act as a platform for developing and testing new methodologies tailored to the conditions and needs of Southeast Asia. Its responsibilities would include preparing manuals and operating procedures, advising on QA/QC activities, coordinating technical audits and visits to monitoring sites, data quality assessments, and storing and facilitating primary calibration standards.

It would also be necessary to establish a broad network of air quality monitoring sites covering regional scales to assess the background levels of atmospheric pollution to which the region is exposed. These stations should not be exposed directly to anthropogenic emissions. They should be located as far away from urban settlements and distinctive emission sources (e.g., roads, factories, power plants, etc.) as possible. In central Peninsular Malaysia, there is currently one monitoring station for this purpose. These stations would allow researchers to assess the influx of pollutants from neighboring regions, as well as the net impact of pollution generated within the region itself as a results of a changing climate. The data from these sites would be essential to evaluate and adjust the results of the regional air quality forecast models, and, in turn, to develop air quality models for operational purposes at the local scale.

In addition to regular air quality monitoring stations, it is common for large cities and areas with air pollution problems to have monitoring supersites. Air quality monitoring supersites are equipped with state-of-the-art instrumentation to provide continuous measurements with higher time-resolution, lower detection limits, and a larger number of chemical species and physical parameters than typical monitoring stations for air quality compliance (e.g., Chazeau et al. 2021; Gani et al. 2020; Zhang et al. 2020; Solomon and Sioutas 2012; see Fig. 10.1). Running a supersite requires skilled personnel to operate and maintain the equipment. Initially, the responsibility could be assigned to a research institution while air quality technicians are trained.

Countries would have to ensure a sufficient budget for operation and maintenance, equipment renewal, and personnel training. It should not be regarded as an expense, but rather as an investment in the protection of public health and the environmental sustainability of cities, natural ecosystems, and agricultural extensions. It will result in a reduction in economic loss due to poor air quality.

Caution must be exercised when using low-cost monitors. Indeed, they are considerably less expensive than regulatory-grade monitors and appear to be much easier to use. They cannot, however, always collect consistent and reliable data. They are less sensitive, precise, and chemically specific than the monitors currently used in monitoring stations for regulatory and warning purposes. Its indiscriminate use would jeopardize the air quality monitoring objectives. They can only be used under specific conditions to complement regulatory monitoring systems and develop new applications to better inform on the state of air quality, but only if a robust calibration and validation scheme is implemented to reduce disparity and uncertainties in their readings.

A uniform air quality index throughout Southeast Asia is currently not feasible due to the current state of air quality management and disparities in air quality monitoring across the region. Furthermore, the characteristics and composition of air pollution vary by location, so air quality indexes must be developed in accordance with background air pollution levels, climatological and topographic conditions, and natural and anthropogenic factors. Instead of a uniform air quality index, the air quality guideline levels and interim targets recommended by the World Health Organization could be used as a means to assess progress toward clean air using

Fig. 10.1 Super air quality monitoring stations are equipped with regulatory-grade sensors to measure criteria pollutants and research-grade instrumentation to continuously measure aerosol chemical composition (e.g., Aerosol Chemical Speciation Monitor, ACSM, Aerodyne Research Inc., Billerica, MA, USA; and Monitor for AeRosols and Gases, MARGA, Metrohm Applikon B. V., Netherlands), black carbon (e.g., Photoacoustic Extinctiometer, PAX, Droplet Measurement Technologies, Boulder, CO, USA; and Dual-spot Aethalometer® AE33, Magee Scientific, Berkeley, CA, USA), and volatile organic compounds (e.g., Gas Chromatography with flame-ionization detector, Perkin Elmer GC-FID system ATX 580, Walthman, MA, USA), among other chemical species. This type of stations are also equipped with advanced instrumentation to measure different meteorological parameters related with the vertical structure of the atmosphere, including fluxes of solar radiation, vertical profiles of temperature and relative humidity (e.g., microwave radiometer profiler, MP-3000A, Radiometrics Corporation, Boulder, USA), and profiles of aerosol backscatter (e.g., Mini Micro Pulse Lidar, Sigma Space Corporation, Lanham, Maryland, USA). The instruments listed here are those installed at the Air Quality Monitoring Supersite of Mexico City, Mexico (http://www.aire.cdmx.gob.mx/)

publicly available air quality data as a reference (see Table 5.1). A metric could be the number of days in a year when air quality levels exceed a chosen interim target. For example, a first step to improve air quality could be to design a series of measures and policies to reduce the number of days that exceed the interim target 3 for $PM_{2.5}$ (37.5 mg/m^3 @ 24-h average) by 10–20% each year, taking into account that current air quality standards in the countries of the region that have them fall between interim targets 1 and 3 (interim targets 1 and 4 are the first and last set of thresholds recommended before achieving the recommended air quality levels, respectively).

However, cities with robust air quality monitoring systems and reliable health statistics could develop and implement health-based air quality indices to better inform their citizens about potential health risks associated with current and previous days' air quality conditions, as well as provide recommendations to reduce exposure to polluted air (e.g., Government of Canada 2021; Government of Hong Kong 2013). Health-based indices are based on local associations between air pollution and health effects, considering hospitalization data. These indices take into account the effects on morbidity and mortality of multiple criteria pollutants and tell more about the health risk posed by air pollution in the short term (e.g., Chen et al. 2023; Perlmutt and Cromar 2019; Stieb et al. 2008). So far in Southeast Asia, such an index has only been developed in Bangkok, but it has not been implemented by authorities (Onchang et al. 2022). Traditional air quality indices provide information on the quality of the air compared to air quality standards based on ambient concentrations of individual pollutants. Notwithstanding, health-based indices cannot replace the traditional indices because the latter trigger control actions to reduce emissions, and provide guidance for reductions in exposure when air quality exceeds specific thresholds. Instead, health-based air quality indexes should be used to guide citizens to make informed decisions on when to modify their outdoor activities and reduce their exposure to air pollution.

Finally, governments would gain credibility and public support for their actions if they made air quality data fully transparent and usable by sharing both time-resolved hourly concentrations and rolling average concentrations, in addition to corresponding air quality indices, in a timely (i.e., near real-time) and analysis-ready format (i.e., easy to analyze). Sharing information about monitors and the locations of monitoring stations would also help to build trust. Environmental agencies are encouraged to share as much information as possible about air quality monitoring activities, including historical records, calibration and operation settings, and QA/QC procedures.

It is critical that citizens have access to detailed information about the quality of the air they breathe. It is the first step toward individual and collective actions. Citizens would become part of the solution by learning on the magnitude of the problem and what would happen if nothing was done to address it. The data would empower people to innovate and propose effective solutions tailored to their specific needs. Full data openness would support environmental journalism and policy advocacy, as well as scientific research and education. Teachers could teach students how to read and analyze air quality records, and citizens who wanted to run their own low-cost monitors would have data to calibrate them.

10.2   Emission Inventories

Emission inventories enable emission mitigation by providing information on what, where, and how emissions are changing. They are necessary to determine the origin of pollutants and precursor species, as well as the quantity emitted by both large and minor emission sources grouped into individual sectors. To take informed action, spatially and temporally distributed emissions data by chemical species are required; however, the construction of emission inventories with such characteristics is still a work in progress in Southeast Asia. No country has an official gridded emissions inventory that can be used for advanced analysis or air quality modeling. Only four countries have compiled lists of annual emission estimates at the national level that include at least three pollutant species and contributions from major emission sectors. In Southeast Asia, the task of building gridded emission inventories has relied on local researchers and international consortia dedicated to building global and regional emission datasets (see Table 6.1). Global and regional emission inventories fill the gap in most developing countries where comprehensive bottom-up emission datasets are lacking. There are a few emission datasets for Southeast Asia that can be used as a starting point for management and research (e.g., EDGAR, CEDS, REAS, ECLIPSE). These are gridded emission inventories built using bottom-up approaches (i.e., emission factors and activity data) at spatial scales from $0.1° \times 0.1°$ to $0.5° \times 0.5°$ (approx. 10–50 km per side) and temporal scales from one hour to one year. They can be used to run regional air quality models, but not at the local or city scale. Users should be aware that there are differences in the emissions estimated by these inventories (see Fig. 6.1), and that emission estimates may not be fully accurate due to the use of generic emission factors and a lack of detailed data for some economic activities.

In light of this, efforts should be directed toward improving the accuracy of emission estimates derived from currently available regional emission inventories. Forming working groups with representatives from each Southeast Asian nation and the lead authors of the global and regional emission inventories listed above, as well as local researchers who have built emission inventories at smaller scales for academic purposes, would be helpful. They should be able to draw a roadmap together to overcome the challenges they have faced when compiling the necessary information to build emission inventories.

It would be useful to develop methodologies tailored to the conditions of Southeast Asia. The proposed methods should act as guidelines for harmonizing the collection of data on economic activities associated with emissions in each country within a regional geographic database. These guidelines should specify the data characteristics and level of detail required to estimate emissions from each economic sector within a certain level of uncertainty. In the event that such a degree of detail, or even the data itself, was not available, they should provide alternative statistical approaches based on proxy variables of said emission source.

Similarly, a common database of pollutant-specific emission factors would help in harmonizing emission estimates across the region. It would be necessary to identify in which sectors the emission factors available in the literature are little or not at all representative of conditions in Southeast Asia. In such cases, the aforementioned working groups should convene experts on the subject to jointly develop missing emission factors through field measurements or laboratory experiments.

Once an updated emissions inventory was available based on information provided by the eleven countries of the region, an important step toward improving air quality management in Southeast Asia would have been taken. This regional emissions inventory, including documentation describing methodologies and listing emission factors, should be fully accessible to the local and international communities. It would act as a reference for global air quality research and assessment. Most importantly, it would offer a platform for the development of emission inventories with much finer spatial resolution (0.5–5 km), allowing them to be used for operational purposes at the scale of provinces or cities in support of the design of air quality improvement programs. These emission inventories would have to be developed by local working groups in close collaboration with environmental authorities and government entities in charge of collecting the necessary data for their preparation.

Regional and local emission inventories must include the systematic application of quality control methods based on verification and statistical analysis of the information used for their preparation. This would allow for coherence, comparability, representativeness, and transparency in the estimated emissions. Similarly, by incorporating quality control processes, the main areas of uncertainty in successive versions of the emission inventories could be identified and resolved.

Similarly, the accuracy of emission inventories must be assessed using independent methods to gain insight into the uncertainties associated with the use of bottom-up approaches. These include data gathering from field observations, modeling activities and air quality monitoring, as well as direct sampling from major emission sources (e.g., stack sampling from large factories, and sampling in tunnels for vehicular emissions), the application of mass balance models and inverse modeling techniques, the use of remote sensing and satellite information processing, and the deployment of eddy covariance flux towers to monitor emissions at neighborhood scale.

Finally, building emission inventories is a continuous task. Emission factors and activity data must be updated on a regular basis to account for ongoing changes in technology, emerging emission sources, the introduction of new emission control measures and environmental policies, changes in people's activities and behaviors, and the effects of climate change. As a result, authorities must provide resources for the work and studies required to update the information used in the inventories.

10.3 Air Quality Modeling and Forecasting

Atmospheric chemical transport models (i.e., air quality models) can explain how pollutants are transported, transformed, and dispersed in the atmosphere. They are useful to analyze past air pollution episodes and future scenarios that take into account changes in emissions and climate, as well as to evaluate the effectiveness of emission mitigation strategies. A well-tuned air quality model can be run daily to forecast air quality conditions in the coming days and thus inform the public about potentially harmful levels of atmospheric pollution. With air quality forecasts, authorities can take precautionary measures, such as reducing industrial activity and restricting vehicle circulation, as well as limiting outdoor activities, to reduce people's exposure to the harmful levels of air pollution expected. However, despite the valuable information that air quality models can provide, no Southeast Asian country has implemented a modeling system as part of its air quality management. This is due to inadequate air quality monitoring and emissions inventory capability, as well as limited availability of qualified technical personnel engaged in air quality modeling.

Air quality models for Southeast Asia have been run by groups of local and international researchers (see Table 7.1). They have run a variety of air quality models (Eulerian, Lagrangian, and hybrid) to answer specific questions about the impact of specific emission sources, such as biomass burning, on individual subregions or the entire region. The majority of these modeling studies have used gridded emission data from global and regional emission datasets. Only a few studies have run models at the province or city scale, having to build their own emissions inventory. To a greater or lesser extent, all studies have validated their modeling outputs using air quality data from regulatory monitoring stations and satellite observations.

The implementation of local and regional air quality operating models is likely the most difficult technical challenge facing air quality management in Southeast Asia. Coordination of efforts would be required between authorities and researchers to move and customize air quality models developed for research purposes to platforms where authorities could use them to support management activities. In the beginning, these models should be used to define needs in the design of monitoring systems and to assist in the preparation of emission inventories. This would require allocating resources for the acquisition of computer infrastructure as well as technical personnel training.

The eleven countries of Southeast Asia should work together to develop the region's modeling capabilities by assembling a suite of models that includes those already available as well as future developments aimed at producing a forecast model that covers the entire region. Then, based on their specific needs, each country, province, and city could use this modeling platform to develop their own modeling capabilities.

Global forecasts for atmospheric composition produced by CAMS, NASA's GEO-CF system, NCAR's WACCM, SILAM, or any other air quality forecasting service could serve as an initial platform for such an endeavor. In Europe, for example, CAMS products are accompanied by coherent air quality multisystem analysis and models to download CAMS regional forecasts at higher spatial resolution for practical use for air quality regulatory and warning purposes (Peuch et al. 2022). This would require the implementation of chemical transport models tailored to local conditions while using the initial and boundary conditions of the global modeling system as input data. Forecasts should be generated in real-time and used to feed directly or indirectly (through postprocessing) a variety of routine time-critical applications for public entities running, for example, official air quality information systems. Taking it a step further, the same outputs could be used to help policymakers' communicate with the general public. One example is the online CAMS Air Control Toolbox, a policy product that allows users to interactively explore mitigation scenarios in air quality forecasts across Europe (Colette et al. 2022; https://policy.atmosphere.cop ernicus.eu/act.php). On a daily basis, users can simulate emission reductions from four activity sectors (traffic, industry, residential heating, and agriculture) and see what quantitative impact these would have on the air quality forecast for the next few days. Tools like this would assist authorities in communicating the rationale behind control measures to reduce air pollution.

The implementation and development of air quality modeling systems in support of air quality management is a continuous process. That is why countries must provide resources to ensure continuous improvement and guarantee their continued existence. It must include resources for developing infrastructure (i.e., software and hardware), as well as increasing the number of experts (programmers, meteorologists, atmospheric scientists, etc.) engaged to improving model performance and developing new applications.

10.4 Satellite Observations

Air quality observation capabilities from space have greatly improved over the last few years. They are opening up a new avenue for air quality management, allowing air quality managers to better monitor air pollution locally, regionally, and across the continent. Nonetheless, the integration of such cutting-edge technologies into air quality management in Southeast Asia is still in its early stages, but could be accelerated given recent rapid advances in satellite technology. At the moment, the SERVIR Southeast Asia Air Quality Monitoring Service (https://aq-tracker-servir. adpc.net/) and the ASMC Regional Haze Situation (https://asmc.asean.org/home) are the only regional air quality products based on satellite observations. From an academic standpoint, a number of studies have used satellite products to assess the distribution of pollutants across the region, and the impact of specific events, such as the smoke-haze produced by biomass burning.

Without making excuses, satellite resources are still largely underutilized everywhere, even in countries that have built and launched satellite instruments to monitor atmospheric pollution. As Prados et al. (2021) explain, air quality managers are not trained to handle satellite products, and there is a lack of trust in the data. Vertical column densities (VCD) of trace gases, and aerosol optical depths (AOD) for aerosols, obtained from satellite retrievals can be difficult to interpret and may not always be an accurate representation of air pollution at the surface level. Indeed, O_3, NO_2, SO_2, and $PM_{2.5}$ are criteria pollutants whose concentrations at surface level can be derived from satellite observations; however, assimilation systems to downscale the VCD and AOD provided by satellite products are not always available. To this, we must add, as previously mentioned, that the limited number of reference-grade monitors at ground level throughout the region jeopardizes the assimilation of satellite data in order to derive representative and accurate concentrations at surface level.

For comparison purposes, how useful would weather monitoring based only on satellite observations be? Imagine we did not have tens of thousands of automated weather stations and radars tuning in to weather models in near real time, so that weather predictions would only be based on satellite data. Indeed, this approach would yield information, but with such large uncertainties that it would be useless for any practical purposes. There would be no weather information available for daily forecasting and severe weather warning systems, streamflow forecasting would be inaccurate, affecting water management and flood warning, flight routes and ocean navigation routes would be disrupted, and so on. A similar situation arises with air quality monitoring in regions where surface monitoring is insufficient. While remote sensing is a powerful tool for covering large areas, it must be trained and validated with direct ground measurements before it can be used for local practical decision-making. In this respect, it is worth highlighting the need to coordinate satellite observations with ground-based observations.

As long as air quality monitoring capabilities in Southeast Asia are not improved, the use of satellite products to support air quality management will be limited. In the case of aerosols, it is also necessary to investigate their chemical and optical characteristics in order to retrieve representative AOD. Given the wide variety of aerosols in the region, this is not an easy task. Southeast Asia is impacted by a myriad of emission sources of biogenic and anthropogenic particles, not to mention the production of secondary aerosols.

Despite current deficiencies in air quality monitoring, environmental agencies could already use satellite images provided by space agencies for public outreach and analysis of major pollution episodes. Satellite images, such as those showing plumes of smoke from widespread wildfires or polluted cities, are compelling and resemble observations made with the human eye or a traditional camera. Air quality managers can access these images through the web applications listed in Chapter 8. These applications update their data products within a few hours of observation, and many of them store at least a decade of data. NASA's Giovanni application is a useful and simple tool that allows users to rapidly visualize, interact with, and analyze air quality data from satellite products without having to download the data (https://giovanni.gsfc.nasa.gov/giovanni/).

Southeast Asia countries should take advantage of the fact that GEMS, the first geostationary satellite mission designed to support air quality management in Asia, almost completely covers the region. GEMS resolves the abundance of criteria pollutants and some reactive trace gases at a scale of a few square kilometers on an hourly basis. This enables the tracking of air pollution patterns in urban and suburban areas during peak hours, the transport of plumes from large individual emission sources (e.g., coal-burning power plants) and biomass burning, and the production of O_3 and other secondary pollutants downwind. The GEMS data stream enables researchers to improve emission inventories and air quality forecasting, monitor population exposure with greater precision, and assess the effectiveness of emission control measures.

Finally, both air quality managers and researchers in Southeast Asia must work closely with individuals affiliated with space agencies in charge of satellite missions to maximize the use of air quality observations from space.

10.5   Climate Change and Air Quality Connections

Climate change can have an impact on air quality, and vice versa, there are multiple linkages between both. In one hand, actions to reduce greenhouse gas (GHG) emissions can reduce the emission of pollutant species, resulting in co-benefits for air quality and climate change (West et al. 2013). For example, by reducing the emission of pollutants such as CO, NO_x, and $PM_{2.5}$ resulting from the combustion of fossil fuels, the emissions of CO_2, the dominant GHG, are also reduced. While on the other hand, by reducing the emission of $PM_{2.5}$ may warm or cool the atmosphere. Particles interact with radiation, forcing climate change. Some particles, such as black carbon, warm by absorbing sunlight, while others, such as sulfates, cool by scattering sunlight and interacts with the clouds, inducing changes in precipitation and regional circulation patterns.

Climate change, as explained by Fiore et al. (2015), is expected to degrade air quality in many polluted areas due to changes in local and regional meteorology, which may have a severe impact on ventilation and dilution, physical processes that modulate the formation and accumulation of pollutants, as well as changes in precipitation and other processes that remove pollutants from the atmosphere. Furthermore, climate change may exacerbate air pollution by triggering some amplifying responses in atmospheric chemistry and emission sources. A warmer atmosphere, for example, enhances O_3 production through an increase in biogenic and evaporative emissions, a strengthening of photochemical reaction rates, and meteorological changes. This is a major issue for many countries that have been unable to reduce O_3 pollution despite stringent emission control measures on precursor species (i.e., VOCs and NO_x). Unfortunately, model projections for alternative climate and air quality scenarios suggest that this problem will worsen. For example, a recent study that combined O_3 data from a chemical transport model and field observations with a machine learning model revealed that under various climate scenarios, O_3 levels in Southeast Asia could increase by 5–20% by the end of the century (Li et al. 2023).

Airborne particles present a conundrum between climate change and air quality. Reductions in SO_2 constrain the formation of sulfate particles, which improves air quality, but it increases near-term warming by unmasking the warming induced by rising CO_2. However, control measures on emission sources with high black carbon to organic carbon ratios, on the other hand, may offset some of the warming induced by SO_2 emission reductions while improving air quality. As already mentioned, black carbon is a major component of particle pollution; it is produced by the incomplete combustion of fossil fuels, biofuels, and biomass, and its presence is always accompanied by sulfates and organic aerosols.

Indeed, control strategies to reduce O_3 and particle pollution do contribute to mitigate climate change, but only partially and in the short term. Both O_3 and black carbon are classified as short-lived climate pollutants (SLCP). These are pollutant species that act as climate forcers, but that have relatively short atmospheric lifetime (days to a few years) when compared to CO_2 (a few centuries and maybe thousands of years). The main SLCPs are black carbon, CH_4, O_3, and hydrofluorocarbons. Despite their short atmospheric lifetimes, they have a high global warming potential and are also listed as pollutants. Measures to reduce their emissions will therefore benefit both air quality and climate change. However, as Pierrehumbert (2014) explains, the effect on climate change mitigation is only temporary since what matters is the cumulative warming effect of long-term CO_2 emissions. Because their impact on climate forcing is not permanent and is spatially and seasonally heterogeneous, reducing SLCP emissions would only have a significant impact as long as no real measures were put in place to bring carbon emissions to zero, which is, of course, not what is required to meet climate change projection targets.

According to this warning, Southeast Asia countries will have to weigh the co-benefits of lowering CO_2 and SLCP emissions without overestimating the positive effects on climate change. The warming and cooling effects of reducing emissions of black carbon and SO_2 must be considered when analyzing the net effect of mitigation strategies. Ruminant livestock, rice cultivation, microbial waste processing (landfills, manure, and wastewater), coal mining, and oil and natural gas systems all emit anthropogenic CH_4. It has a 26-fold higher radiative efficiency than CO_2 but a much shorter lifetime of about 12 years. Although CH_4 is not toxic in and of itself, its role in atmospheric chemistry makes it important for air quality management. As explained above, tropospheric O_3 is not a pollutant that is directly emitted into the atmosphere; it is a secondary pollutant formed by atmospheric photochemical processes that must be controlled by reducing its precursor species, primarily NO_x, CO, and VOCs, as well as CH_4.

A better understanding of the impacts of climate change on meteorology, human health, and the environment in Southeast Asia is urgently needed. Climate patterns such as the *El Niño* Southern Oscillation (ENSO) and the Indian Ocean Dipole (IOD) must be evaluated in terms of changes in regional meteorology and their impact on air quality. When ENSO and IOD coincide, for example, anomalous dry conditions occur along the Strait of Malacca, triggering wildfires in Sumatra, Kalimantan, and Peninsular Malaysia, resulting in drastic deterioration in air quality (Field et al. 2009). With the current trend of climate change, this phenomenon appears to be more frequent, and environmental managers must consider its impact on their decisions to protect air quality.

Finally, since many GHG and pollutants species are released from the same emission sources, integrating emission inventories and climate change mitigation strategies with air quality programs would be convenient. This would help in quantifying the health and economic benefits of reducing emissions of both air pollutants and GHGs at the same time. To develop effective policies, air quality managers and researchers would need to develop evidence-based knowledge to inform the public about such benefits, thereby involving them in the mitigation strategies.

10.6 Scientific Research

Scientific research is at the center of the air quality management cycle (see Fig. 1.2). Each component of the cycle must be supported by scientific research to provide solid foundations for designing effective control measures. Air quality managers must have a thorough understanding of the origin, transformation, dispersion, and fate of all atmospheric pollutants, as well as their impact on human health and the ecosystem, in order to take informed action.

Local efforts have focused on better understanding the patterns, trends, and impact of particle pollution caused by agricultural waste burning in the mainland region and peatland fires in the maritime region and Peninsular Malaysia. Several studies have investigated the chemical composition and physical properties of the particles, as well as their origin and adverse effects on public health through source apportionment and epidemiological studies, respectively. A growing number of studies have also focused on other aspects of air quality in large urban areas such as Manila, Bangkok, Hanoi, Ho Chi Minh City, Kuala Lumpur, and Singapore, but very little research has been done outside of these cities.

There are also a number of topics relevant to air quality management that have received little or no attention, including: atmospheric chemistry, climate change impacts on air quality and vice versa, boundary layer meteorology, the impact of micrometeorological phenomena on pollutants dispersion, emergent and persistent pollutants, toxicology, and ecosystem damage. Similarly, there is a need to develop monitoring and analysis methods that are tailored to the needs and conditions of Southeast Asia, as well as chemical transport models and air quality multisystem analysis to download regional air quality forecasts and satellite products that can be used for air quality regulatory and warning purposes.

As already mentioned, it is the academic community that has taken on the task of building emission inventories and running air quality models in Southeast Asia (see Tables 6.1 and 7.1). Although local researchers have made significant and valuable contributions, the most relevant contributions on these two issues have come from the international community. Some of the initiatives from which these have emerged have involved local researchers and, even better, graduate students, but the current academic system does not appear to encourage local researchers to continue such efforts to the same or greater extent.

The same is true for large field studies. Local scientists and students are often involved, but they do not lead such studies and their input is sometimes limited. It is common for international initiatives to include local researchers to comply with funding agency policies, but without seeking substantive input from them. These academic initiatives are known as parachute science, and they are distinguished by the fact that they produce no intellectual exchange and frequently serve only the needs of foreign researchers. However, many institutions and local researchers accept them because of the funding and prestige that comes with international partnership.

A good example against this pernicious practice comes from the Manila Observatory, whose researchers have established collaborations with overseas institutions to carry out field studies in which they have co-led the work alongside their international peers. This has enabled them to get access to state-of-the-art instrumentation, gain the experience of researchers from other latitudes, and most importantly, conduct cutting-edge research.

Collaborations between scientists in the region are uncommon. Those that do exist appear to respond more to personal efforts than institutional efforts. This is an issue that universities and research institutions throughout the region must address for the sake of their societies. Researchers must see the value in working together, taking on the scientific needs of their neighbors as their own within a framework of "collective advantage in shared knowledge."

In the same context, until now, environmental agencies' participation in scientific studies has been limited. It is critical that air quality managers and researchers work together to propose and carry out joint studies that generate the scientific information needed for decision-making.

Many countries have learned that atmospheric phenomena can jeopardize their economic and social development, so what happens in the atmosphere is considered a matter of national security. In consequence they have invested in cutting-edge research centers to better understand the physical and chemical processes related to the atmospheric environment in order to be prepared for severe weather conditions and meteorological anomalies caused by a changing climate, as well as to mitigate the impacts of air pollution, GHG emissions, and global warming. For example, research institutes such as the *Norwegian Institute for Air Research* (https://www.nilu.no/), the *Swedish Meteorological and Hydrological Institute* (https://www.smhi.se/), the *National Institute for Environmental Studies* (https://www.nies.go.jp/) in Japan, and the *National Center for Atmospheric Research* (https://ncar.ucar.edu/) in the US have been instrumental in providing scientific information to support air quality management in their respective countries. The scientists at these research centers have made significant contributions to the improvements in air quality that their citizens have experienced.

University research should also help to solve air quality problems. However, it must be noted that the university dynamics make meeting environmental management needs difficult. University professors must teach and provide administrative services to the institution in addition to conducting research. The scientific research index published by the journal Nature reflects this. It highlights that the participation of university researchers in high-impact research work in environmental sciences is lower than in other areas of knowledge, whereas the scientific contribution of government institutions is higher (Armitage 2018).

In this regard, it is recommended that the eleven countries of Southeast Asia establish a joint research center aimed at generating scientific information to supports air quality management at the regional level. This information will allow air quality managers to address and resolve environmental problems in a collaborative and timely manner, truly serving the needs of Southeast Asia without the need for external consultants and foreign research institutions.

Large collaborative research initiatives should ideally come from local institutions rather than from overseas institutions, as has frequently occurred. Any research initiative is welcome, but those that are nurtured locally are more likely to better address the needs of the region, based on the assumption that local researchers are better aware of their own needs as well as the regional situation than foreign researchers. Furthermore, locally proposed studies are more likely to be continued. Studies proposed and coordinated by foreign researchers generate valuable information and draw the attention of the international community to the region, but they rarely go further, and their local impact fades. In this context, it would be wise for local researchers to take the initiative and invite the international community to collaborate on their projects, identifying the research questions to be addressed together. Local researchers would gain experience from their peers, as well as access to equipment that they might not be able to afford otherwise. International researchers, for their part, would generate knowledge and information that would assist them in solving environmental problems in their communities, given that air pollution has no borders.

10.7   Collaborative Air Quality Field Campaigns

Collaborative intensive field campaigns are an effective way to promote scientific research. International experience indicates that they are a means for collecting air quality data to improve scientific knowledge and address air pollution (Velasco et al. 2021). They typically involve the deployment of a wide range of research-grade monitors, often including state-of-the-art instrumentation, as well as remote sensing devices across networks of ground observation sites and tall towers, mobile laboratories, and onboard research aircrafts, with the assistance of satellite observations and meteorological and chemical forecasting models. The synergy of using multiple analytical techniques mounted on different measurement platforms in conjunction with remote sensing data and meteorological observations at different scales can improve the knowledge of the chemistry, dispersion, and transport processes of primary and secondary pollutants, as well as provide the necessary information to advance the development of emission inventories and the performance of air quality models. Extensive data on the chemical composition of the atmosphere, as well as key meteorological parameters, help to reduce uncertainties in emission estimates and numerical simulations, and provide guidance for setting priorities for improving air quality management.

Only a small number of these type of studies have been conducted in Southeast Asia to investigate the impacts of biomass burning on the region's ambient air quality. For example, the collaborative field campaigns Biomass-burning Aerosols in South-East Asia: Smoke Impact Assessment (BASE-ASIA) in 2006 and the 7-South-East Asian Studies (7-SEAS) in 2010–2013 integrated a mix of models, emission inventories, and *in-situ* and remote observational data to answer questions about the type, frequency, intensity, and geographical extent of impacts caused by biomass burning on regional air quality, radiative forcing, cloud properties, and weather (Huang et al. 2013; Reid et al. 2013, 2015; Tsay et al. 2013).

The Clouds, Aerosol Monsoon Processes Philippines Experiment (CAMP2Ex) was a NASA-led collaborative field campaign to understand the interdisciplinary relationships between aerosol lifecycle, convection, and the radiation budget within the Southeast Asian southwest monsoon system. It was a multi-year monitoring effort that included the operation of an aerosol and radiation monitoring site at the Manila Observatory. CAMP2Ex was centered on a 42-day (Aug–Oct 2019) airborne campaign in and around the Philippines, which included two research aircraft outfitted with radars, microwave sounders, lidar, radiation, cloud microphysics, composition, and state instrumentation, aligned with significant remote sensing data collections and informatics efforts. Although its main objective was not to collect data to improve air quality management, some of the measurements have been used to address issues related to particle pollution in Manila (e.g., Lorenzo et al. 2021; Braun et al. 2020; Stahl et al. 2020; Cruz et al. 2019).

Another collaborative field campaign worth mentioning is the Manila Aerosol Characterization Experiment, MACE-2015, in which a consortium of Philippine researchers collaborated with scientists from the Leibniz Institute for Tropospheric Research to characterize traffic particles and understand their environmental and public health impacts with the aim of assisting in the design of targeted solutions (Alas et al. 2018; Kecorius et al. 2017).

The Airborne and Satellite Investigation of Asian Air Quality (ASIA-AQ) cooperative field study was underway at the time of writing as of early 2024. ASIA-AQ aimed to improve the integration of satellite data with existing air quality ground monitoring and modeling activities in Asia (https://espo.nasa.gov/asia-aq/). It revolved around GEMS satellite observations and included short-term airborne measurements from two NASA's instrumented aircrafts over the Philippines and Thailand, focusing on the metropolitan areas of Manila, Bangkok, and Chiang Mai. It was conducted in full partnership with local scientists and environmental agencies. The results are expected to provide chemical details and vertically resolved information on the distribution of pollutants throughout the overlying atmospheres of these cities, as well as better characterizing their emissions and identifying major sources, and assessing processes leading to secondary production of O_3 and fine particles.

Air quality management in Southeast Asia would benefit from periodic intensive air quality field campaigns like those mentioned here and as it is done in other parts of the world through international collaboration. These campaigns should focus on researching and generating information on the factors and processes that drive air pollution at different scales, from regional to city scale, with the aim of closing knowledge gaps that hinder further progress in air quality management. Approaches and strategies can differ depending on the needs and air pollution characteristics. Given the region's limited infrastructure and limited data on air quality, it would be reasonable to focus first on understanding the main causes, dynamics, and magnitude of the problem. This would allow for the implementation of initial control measures for capping major emission sources while investing in monitoring infrastructure, developing emission inventories, and nurturing air quality modeling capabilities to assess the effectiveness of such measures. For later stages, when the region already has solid air quality management programs, well-instrumented air quality monitoring infrastructure, robust emission inventories, and reliable modeling systems, intensive field campaigns could focus on detailed characterizations of the physical and chemical processes behind the production of secondary pollutants, including testing of new measurement techniques and chemical mechanisms, as well as novel studies to evaluate health impacts and economic losses.

Not only do collaborative intensive field campaigns provide science-based information to support the design and updating of air quality management policies, but they also promote technological advances in air quality monitoring and measurements. They enable the testing of novel measurement techniques as well as the performance of new monitors that can be integrated into air quality monitoring networks in future. Furthermore, intensive field campaigns allow for the training of technical personnel. These studies have the potential to act as catalysts for increased institutional capacity building.

Intensive air quality field campaigns are highly demanding initiatives in terms of financial and human resources. It takes one or two years to design, secure funds, prepare for, and execute the field measurements. From one to three months of intensive measurements must be considered to collect enough data to address the variability and representativity of the samplings. The combined data analysis and numerical modeling can take at least two years; thus, the entire endeavor will take four to five years to complete. The resources needed for these studies depend on their scope and scale, and since they must be covered up-front, they can appear relatively high, especially if environmental agencies are not adequately funded. However, their cost represents a small percentage of the economic cost associated with air pollution.

Authorities and decision-makers in charge of air quality management and regulations should be involved in the design, implementation, and analysis of the field campaign. The key air quality challenges must be discussed with them in order to identify the scientific knowledge needed to develop or update air quality policies. Their participation would help the study results lead to direct decisions. Furthermore, the participation of authorities would facilitate the practical aspects of the study. For example, they would assist in obtaining the necessary permits to deploy instrumentation, build observation platforms, drive mobile laboratories, and fly instrumented aircrafts, as well as get access to complementary data on diverse aspects for the results analysis and modeling.

References

Alas, H.D., Müller, T., Birmili, W., Kecorius, S., Cambaliza, M.O., Simpas, J.B.B., Cayetano, M., Weinhold, K., Vallar, E., Galvez, M.C., Wiedensohler, A.: Spatial characterization of black carbon mass concentration in the atmosphere of a Southeast Asian megacity: an air quality case study for Metro Manila, Philippines. Aerosol Air Quality Res. **18**(9), 2301–2317 (2018). https://doi.org/10.4209/aaqr.2017.08.0281

Armitage, C.: The search for solutions. Nature Index 2018 Earth Environ. Sci. **S1** (2018). https://doi.org/10.1038/d41586-018-05490-6

Braun, R.A., Aghdam, M.A., Bañaga, P.A., Betito, G., Cambaliza, M.O., Cruz, M.T., Lorenzo, G.R., MacDonald, A.B., Simpas, J.B., Stahl, C., Sorooshian, A.: Long-range aerosol transport and impacts on size-resolved aerosol composition in Metro Manila, Philippines. Atmos. Chem. Phys. **20**, 2387–2405 (2020). https://doi.org/10.5194/acp-20-2387-2020

Chazeau, B., Temime-Roussel, B., Gille, G., Mesbah, B., D'Anna, B., Wortham, H., Marchand, N.: Measurement report: Fourteen months of real-time characterization of the submicronic aerosol and its atmospheric dynamics at the Marseille-Longchamp supersite. Atmos. Chem. Phys. **21**, 7293–7319 (2021). https://doi.org/10.5194/acp-21-7293-2021

Chen, M.J., Guo, Y.L., Lin, P., Chiang, H.C., Chen, P.C., Chen, Y.C.: Air quality health index (AQHI) based on multiple air pollutants and mortality risks in Taiwan: construction and validation. Environ. Res. **231**, 116214 (2023)

Colette, A., Rouïl, L., Meleux, F., Lemaire, V., Raux, B.: Air control toolbox (ACT_v1.0): a flexible surrogate model to explore mitigation scenarios in air quality forecasts. Geoscientific Model Develop. **15**, 1441–1465 (2022). https://doi.org/10.5194/gmd-15-1441-2022

Cruz, M.T., Bañaga, P.A., Betito, G., Braun, R.A., Stahl, C., Aghdam, M.A., Cambaliza, M.O., Dadashazar, H., Hilario, M.R., Lorenzo, G.R., Ma, L., MacDonald, A.B., Pabroa, P.C., Yee, J.R., Simpas, J.B., Sorooshian, A.: Size-resolved composition and morphology of particulate matter during the southwest monsoon in Metro Manila, Philippines. Atmos. Chem. Phys. **19**(16), 10675–10696 (2019). https://doi.org/10.5194/acp-19-10675-2019

Field, R.D., van der Werf, G.R., Shen, S.S.P.: Human amplification of drought-induced biomass burning in Indonesia since 1960. Nat. Geosci. **2**, 185–188 (2009). https://doi.org/10.1038/ngeo443

Fiore, A.M., Naik, V., Leibensperger, E.M.: Air quality and climate connections. J. Air Waste Manag. Assoc. **65**(6), 645–685 (2015). https://doi.org/10.1080/10962247.2015.1040526

Gani, S., Bhandari, S., Patel, K., Seraj, S., Soni, P., Arub, Z., Habib, G., Hildebrandt Ruiz, L., Apte, J.S.: Particle number concentrations and size distribution in a polluted megacity: the Delhi Aerosol Supersite study. Atmos. Chem. Phys. **20**, 8533–8549 (2020). https://doi.org/10.5194/acp-20-8533-2020

Gani, S., Pant, P., Sarkar, S., Sharma, N., Dey, S., Guttikunda, S.K., AchutaRao, K.M., Nygard, J., Sagar, A.D.: Systematizing the approach to air quality measurement and analysis in low and middle income countries. Environ. Res. Lett. **17**, 021004 (2022). https://doi.org/10.1088/1748-9326/ac4a9e

Government of Canada. Air Quality Health Index. Environment and Natural Resources https://www.canada.ca/en/environment-climate-change/services/air-quality-health-index.htmll.   Accessed 26 Dec 2023

Government of Hong Kong. Air Quality Health Index. Environmental Protection Department. The Government of the Hong Kong Special Administrative Region. https://www.aqhi.gov.hk/en.html. Accessed 26 Dec 2023

Hasenkopf, C., Sharma, N., Mukerjee, P., Greenstone, M.: The case for closing global air quality data gaps with local actors: a golden opportunity for the philanthropic community. In: Energy Policy Institute at the University of Chicago, Chicago, IL, USA. https://epic.uchicago.edu/ (2023)

Health Effects Institute (HEI). State of Global Air 2024. Special Report. Health Effects Institute, Boston, MA (2024). https://www.stateofglobalair.org/

Hidy, G.M., Brook, J.R., Demerjian, K.L., Molina, L.T., Pennell, W.T., Scheffe, R.D.: Technical Challenges Of Multipollutant Air Quality Management. Springer, Dordrecht, The Netherlands (2011). ISBN: 978-94-007-0304-9. https://doi.org/10.1007/978-94-007-0304-9

Huang, K., Fu, J.S., Hsu, N.C., Gao, Y., Dong, X., Tsay, S.C., Lam, Y.F.: Impact assessment of biomass burning on air quality in Southeast and East Asia during BASE-ASIA. Atmos. Environ. **78**, 291–302 (2013). https://doi.org/10.1016/j.atmosenv.2012.03.048

Institute for Health Metric and Evaluation (IHME). In: Global Burden of Diseases 2019. University of Washington. https://www.healthdata.org/gbd/2019 (2020)

Kecorius, S., Madueño, L., Vallar, E., Alas, H., Betito, G., Birmili, W., Cambaliza, M.O., Catipay, G., Gonzaga-Cayetano, M., Galvez, M.C., Lorenzo, G., Müller, T., Simpas, J.B., Tamayo, E.G., Wiedensohler, A.: Aerosol particle mixing state, refractory particle number size distributions and emission factors in a polluted urban environment: Case study of Metro Manila, Philippines. Atmos. Environ. **170**, 169–183 (2017). https://doi.org/10.1016/j.atmosenv.2017.09.037

Li, H., Yang, Y., Jin, J., Wang, H., Li, K., Wang, P., Liao, H.: Climate-driven deterioration of future ozone pollution in Asia predicted by machine learning with multi-source data. Atmos. Chem. Phys. **23**, 1131–1145 (2023). https://doi.org/10.5194/acp-23-1131-2023

Lorenzo, G.R., Bañaga, P.A., Cambaliza, M.O., Cruz, M.T., AzadiAghdam, M., Arellano, A., Betito, G., Braun, R., Corral, A.F., Dadashazar, H., Edwards, E.-L., Eloranta, E., Holz, R., Leung, G., Ma, L., MacDonald, A.B., Reid, J.S., Simpas, J.B., Stahl, C., Visaga, S.M., Sorooshian, A.: Measurement report: Firework impacts on air quality in Metro Manila, Philippines, during the 2019 New Year revelry. Atmos. Chem. Phys. **21**, 6155–6173 (2021). https://doi.org/10.5194/acp-21-6155-2021

Molina, L.T., Velasco, E., Retama, A., Zavala, M.: Experience from integrated air quality management in the Mexico City Metropolitan Area and Singapore. Atmosphere **10**(9), 512 (2019). https://doi.org/10.3390/atmos10090512

Onchang, R., Hirunkasi, K., Janchay, S.: Establishment of a city-based index to communicate air pollution-related health risks to the public in Bangkok, Thailand. Sustainability **14**(24), 16702 (2022). https://doi.org/10.3390/su142416702

Perlmutt, L.D., Cromar, K.R.: Comparing associations of respiratory risk for the EPA Air Quality Index and health-based air quality indices. Atmos. Environ. **202**, 1–7 (2019). https://doi.org/10.1016/j.atmosenv.2019.01.011

Peuch, V., Engelen, R., Rixen, M., Dee, D., Flemming, J., Suttie, M., Ades, M., Agustí-Panareda, A., Ananasso, C., Andersson, E., Armstrong, D., Barré, J., Bousserez, N., Dominguez, J. J., Garrigues, S., Inness, A., Jones, L., Kipling, Z., Letertre-Danczak, J., Parrington, M., Razinger, M., Ribas, R., Vermoote, S., Yang, X., Simmons, A., Garcés de Marcilla, J., Thépaut, J.: The Copernicus atmosphere monitoring service: from research to operations. Bull. Am. Meteorol. Soc. **103**(12), E2650–E2668 (2022). https://doi.org/10.1175/BAMS-D-21-0314.1

Pierrehumbert, R.T.: Short-lived climate pollution. Annu. Rev. Earth Planet. Sci. **42**, 341–379 (2014). https://doi.org/10.1146/annurev-earth-060313-054843

Pinder, R.W., Klopp, J.M., Kleiman, G., Hagler, G.S., Awe, Y., Terry, S.: Opportunities and challenges for filling the air quality data gap in low- and middle-income countries. Atmos. Environ. **215**, 116794 (2019). https://doi.org/10.1016/j.atmosenv.2019.06.032

Prados, A.I., Vernier, J.P., Duncan, B., Lamsal, L.: The present and future of satellite observations for US air quality management. In: The Magazine for Environmental Managers, Air and Waste Management Association. https://www.awma.org/emcsept21 (2021)

Reid, J.S., Hyer, E.J., Johnson, R.S., Holben, B.N., Yokelson, R.J., Zhang, J., Campbell, J.R., Christopher, S.A., Di Girolamo, L., Giglio, L., Holz, R.E., Kearney, C., Miettinen, J., Reid, E.A., Turk, F.J., Wang, J., Xian, P., Zhao, G., Balasubramanian, R., Chew, B.N., Janjai, S., Lagrosas, N., Lestari, P., Lin, N.-H., Mahmud, M., Nguyen, A.X., Norris, B., Oanh, N.T.K., Oo, M., Salinas, S.V., Welton, E.J., Liew, S.C.: Observing and understanding the Southeast Asian aerosol system by remote sensing: an initial review and analysis for the Seven Southeast Asian Studies (7SEAS) program. Atmos. Res. **122**, 403–468 (2013). https://doi.org/10.1016/j.atmosres.2012.06.005

Reid, J.S., Lagrosas, N.D., Jonsson, H.H., Reid, E.A., Sessions, W.R., Simpas, J.B., Uy, S.N., Boyd, T.J., Atwood, S.A., Blake, D.R., Campbell, J.R., Cliff, S.S., Holben, B.N., Holz, R.E., Hyer, E.J., Lynch, P., Meinardi, S., Posselt, D.J., Richardson, K.A., Salinas, S.V., Smirnov, A., Wang, Q., Yu, L., Zhang, J.: Observations of the temporal variability in aerosol properties and their relationships to meteorology in the summer monsoonal South China Sea/East Sea: the scale-dependent role of monsoonal flows, the Madden–Julian Oscillation, tropical cyclones, squall lines and cold pools. Atmos. Chem. Phys. **15**, 1745–1768 (2015). https://doi.org/10.5194/acp-15-1745-2015

Solomon, P.A., Sioutas, C.: Continuous and semicontinuous monitoring techniques for particulate matter mass and chemical components: A synthesis of findings from EPA's Particulate Matter Supersites Program and related studies. J. Air Waste Manage. Assoc. **58**(2), 164–195 (2012). https://doi.org/10.3155/1047-3289.58.2.164

Stahl, C., Cruz, M.T., Bañaga, P.A., Betito, G., Braun, R.A., Aghdam, M.A., Cambaliza, M.O., Lorenzo, G.R., MacDonald, A.B., Hilario, M.R.A., Pabroa, P.C.: Sources and characteristics of size-resolved particulate organic acids and methanesulfonate in a coastal megacity: Manila, Philippines. Atmos. Chem. Phys. **20**(24), 15907–15935 (2020). https://doi.org/10.5194/acp-20-15907-2020

Stieb, D.M., Burnett, R.T., Smith-Doiron, M., Brion, O., Shin, H.H., Economou, V.: A new multi-pollutant, no-threshold air quality health index based on short-term associations observed in daily time-series analyses. J. Air Waste Manag. Assoc. **58**(3), 435 (2008). https://doi.org/10.3155/1047-3289.58.3.435

Tsay, S.C., Hsu, N.C., Lau, W.K.M., Li, C., Gabriel, P.M., Ji, Q., Holben, B.N., Welton, E.J., Nguyen, A.X., Janjai, S., Lin, N.H., Reid, J.S., Boonjawat, J., Howell, S.G., Huebert, B.J., Fu, J.S., Hansell, R.A., Sayer, A.M., Gautam, R., Wang, S.H., Goodloe, C.S., Miko, L.R., Shu, P.K., Loftus, A.M., Huang, J., Kim, J.Y., Jeong, M.Y., Pantina, P.: From BASE-ASIA toward 7-SEAS: a satellite-surface perspective of boreal spring biomass-burning aerosols and clouds in Southeast Asia. Atmos. Environ. **78**, 20–34 (2013). https://doi.org/10.1016/j.atmosenv.2012.12.013

Velasco, E., Retama, A., Zavala, M., Guevara, M., Rappenglück, B., Molina, L.T.: Intensive field campaigns as a means for improving scientific knowledge to address urban air pollution. Atmos. Environ. **246**, 118094 (2021). https://doi.org/10.1016/j.atmosenv.2020.118094

West, J.J., Smith, S.J., Silva, R.A., Naik, V., Zhang, Y., Adelman, Z., Fry, M.M., Anenberg, S., Horowitz, L.W., Lamarque, J.F.: Co-benefits of mitigating global greenhouse gas emissions for future air quality and human health. Nat. Clim. Change **3**, 885–889 (2013). https://doi.org/10.1038/nclimate2009

World Bank: The global health cost of $PM_{2.5}$ Air pollution: a case for action beyond 2021. In: International Development in Focus. World Bank, Washington, DC. https://openknowledge.worldbank.org/handle/10986/36501 (2022)

World Health Organization (WHO): Improving the capacity of countries to report on air quality in cities: users' guide to the repository of United Nations tools. WHO, Geneva, Switzerland. https://www.who.int/publications/i/item/9789240074446 (2023)

Zhang, C., Zou, Z., Chang, Y., Zhang, Y., Wang, X., Yang, X.: Source assessment of atmospheric fine particulate matter in a Chinese megacity: Insights from long-term, high-time resolution chemical composition measurements from Shanghai flagship monitoring supersite. Chemosphere **251**, 126598 (2020). https://doi.org/10.1016/j.chemosphere.2020.126598

Chapter 11
In Closing

Abstract Delaying the implementation of emission control measures to improve air quality, both locally and regionally, increases the requirements for attaining clean air, the costs associated with breathing unhealthy air, and positive feedback on climate change efforts. There are gaps in air quality management to fill, additional areas to investigate, and environmental policies to improve. Achieving clean air across the region will demonstrate Southeast Asian nations' ability to work together to address current and future threats caused by a changing climate, as well as economic and political obstacles. Southeast Asia does not have to start from scratch. To a greater or lesser extent, its eleven nations have made progress in tackling air pollution, the lessons learned should serve as a starting point. Similarly, the experience acquired by nations in other parts of the world when integrating strategies to improve air quality should be used as a reference. The monograph closes with a set of recommendations on how policymakers, air quality managers, scientists, practitioners, educators, professional communicators and journalists, and the non-governmental community can collaborate to achieve clean air with a local and regional vision.

Keywords Air quality policy · Science-policy approach · Transparency · Governance · Education

Recognizing that each country is both an emitter and a receiver of air pollutants, it is imperative that the eleven Southeast Asian nations work together to fill the gaps in air quality management throughout the region. The knowledge on air pollution is limited and insufficient to take informed action at the local and regional scales. Concerted efforts among air quality managers, policymakers, and the scientific community from all countries are needed to achieve clean air. The authors hope that the recommendations below will help Southeast Asian nations develop a roadmap for improving air quality.

I. National air quality policy frameworks must be scaled up to the regional level. Countries will need to set measurable targets together to reduce air pollution. Specific policy support will be required to incentivize air quality management (i.e., air quality monitoring, emission inventories, and air quality modeling and forecasting) while ensuring it is done correctly through close monitoring of progress and good public governance.

II.   There must be explicit recognition of the gap in air quality management development across the region. Policymakers must understand the costs and consequences of failing to address the threat posed by air pollution. Nations will need to work together to develop the tools needed to collect the data required for policymakers to address air pollution following a science-policy approach. In terms of environmental justice and equity, it is critical to have reliable information on air pollution and associated risks, independent of the country's development level. Under clear transparency rules, air quality managers, and policymakers will have to identify the financial requirements and schemes that allow such an endeavor while taking into account the economic conditions of each country.

III. Southeast Asian countries need concerted efforts to build institutions and communities for collecting and sharing data and information. A larger local community of air quality and atmospheric science experts is required to broaden knowledge, perspectives, and experiences to guide the development of air quality management. Similarly, all countries need to improve the availability of data on air quality projects, plans, investment, and other relevant dimensions.

IV.   Both technical and human resources will be required to hon the analysis around a more complete and consistent perspective of the air quality state throughout the region, as well as to outline short- and long-term steps to achieve clean air. Closing evidence gaps requires local research, especially on topics where data or knowledge is limited or missing. Similarly, adapting available air quality management tools to Southeast Asian conditions will require engineering and applied research efforts.

V.   What is now understood as air quality management will evolve as new methods are developed, research continues, and new technologies emerge. Therefore, sustained investments in infrastructure and trained personnel are required.

VI.   Authorities and society must speed up the implementation of emission control measures through educational outreach. The benefits of breathing clean air should extend beyond English-language scientific literature. Air quality experts, in collaboration with educators, communicators, and journalists, should teach society how to use the information derived from air quality management for their individual and collective benefit. Ultimately, accurate data, good science, and well-chosen technologies can direct the way to corrective regulatory measures, but without strong commitment from the government and civil society, no amount of data, science, or technology will solve the air pollution problem.

Index